D1711290

Lecture Notes in Physics

The Editorial Policy for Monographs

The series Lecture Notes in Physics reports new developments in physical research and teaching - quickly, informally, and at a high level. The type of material considered for publication includes monographs presenting original research or new angles in a classical field. The timeliness of a manuscript is more important than its form, which may be preliminary or tentative. Manuscripts should be reasonably self-contained. They will often present not only results of the author(s) but also related work by other people and will provide sufficient motivation, examples, and applications.

Acceptance

The manuscripts or a detailed description thereof should be submitted either to one of the series editors or to the managing editor. The proposal is then carefully refereed. A final decision concerning publication can often only be made on the basis of the complete manuscript, but otherwise the editors will try to make a preliminary decision as definite as they can on the basis of the available information.

Contractual Aspects

Authors receive jointly 30 complimentary copies of their book. No royalty is paid on Lecture Notes in Physics volumes. But authors are entitled to purchase directly from Springer other books from Springer (excluding Hager and Landolt-Börnstein) at a $33\frac{1}{3}\%$ discount off the list price. Resale of such copies or of free copies is not permitted. Commitment to publish is made by a letter of interest rather than by signing a formal contract. Springer secures the copyright for each volume.

Manuscript Submission

Manuscripts should be no less than 100 and preferably no more than 400 pages in length. Final manuscripts should be in English. They should include a table of contents and an informative introduction accessible also to readers not particularly familiar with the topic treated. Authors are free to use the material in other publications. However, if extensive use is made elsewhere, the publisher should be informed. As a special service, we offer free of charge LaTeX macro packages to format the text according to Springer's quality requirements. We strongly recommend authors to make use of this offer, as the result will be a book of considerably improved technical quality. The books are hardbound, and quality paper appropriate to the needs of the author(s) is used. Publication time is about ten weeks. More than twenty years of experience guarantee authors the best possible service.

LNP Homepage (springerlink.com)

On the LNP homepage you will find:
—The LNP online archive. It contains the full texts (PDF) of all volumes published since 2000. Abstracts, table of contents and prefaces are accessible free of charge to everyone. Information about the availability of printed volumes can be obtained.
—The subscription information. The online archive is free of charge to all subscribers of the printed volumes.
—The editorial contacts, with respect to both scientific and technical matters.
—The author's / editor's instructions.

G. Cassinelli, E. De Vito, P. J. Lahti, A. Levrero

The Theory of Symmetry Actions in Quantum Mechanics

with an Application to the Galilei Group

Authors

Gianni Cassinelli
Universita di Genova
Dipartimento di Fisica
16146 Genova, Italy

Ernesto De Vito
Dipartimento di Matematica
Pura ed Applicata "G. Vitali"
41000 Modena, Italy

Pekka J. Lahti
University of Turku
Department of Physics
20014 Turku, Finland

Alberto Levrero
Universita di Genova
Dipartimento di Fisica
16146 Genova, Italy

G. Cassinelli, E. De Vito, P. J. Lahti, A. Levrero, *The Theory of Symmetry Actions in Quantum Mechanics*, Lect. Notes Phys. **654** (Springer, Berlin Heidelberg 2004), DOI 10.1007/b99455

Library of Congress Control Number: 2004110193

ISSN 0075-8450
ISBN 3-540-22802-0 Springer Berlin Heidelberg New York

Springer is a part of Springer Science+Business Media

springeronline.com

Typesetting: Camera-ready by the authors/editor
Data conversion: PTP-Berlin Protago-TeX-Production GmbH
Cover design: *design & production*, Heidelberg

Printed on acid-free paper
54/3141/ts - 5 4 3 2 1 0

This book is dedicated to Enrico Beltrametti
on the occasion of his seventieth birthday.

Preface

This book is devoted to the study of the symmetries in quantum mechanics. In many elementary expositions of quantum theory, one of the basic assumptions is that a group G of transformations is a group of symmetries for a quantum system if G admits a unitary representation U acting on the Hilbert space $\mathcal{H}$ associated with the system. The requirement that, given $g \in G$, the corresponding operator U_g is unitary is motived by the need for preserving the transition probability between any two vector states $\varphi, \psi \in \mathcal{H}$,

$$|\langle \varphi, U_g \psi \rangle|^2 = |\langle \varphi, \psi \rangle|^2. \tag{0.1}$$

The composition law

$$U_{g_1 g_2} = U_{g_1} U_{g_2} \tag{0.2}$$

encodes the assumption that the physical symmetries form a group of transformations on the set of vector states.

However, as soon as one considers some explicit application, the above framework appears too restrictive. For example, it is well known that the wave function φ of an electron changes its sign under a rotation of 2π; the Dirac equation is not invariant under the Poincaré group, but under its universal covering group; the Schrödinger equation is invariant neither under the Galilei group nor under its universal covering group.

The above *pathologies* have important physical consequences: bosons and fermions can not be coherently superposed, the canonical position and momentum observables of a Galilei invariant particle do not commute and particles with different mass cannot be coherently superposed.

For the Poincaré group the above problem was first solved by Wigner in his seminal paper [40] and it was systematically studied by Bargmann, [1], and Mackey, [27] (see, also, the book of Varadarajan, [35], for a detailed exposition of the theory).

These authors clarified that in order to preserve (0.1), one only has to require that U be either unitary or antiunitary and (0.2) can be replaced by the weaker condition

$$U_{g_1 g_2} = m(g_1, g_2) U_{g_1} U_{g_2}, \tag{0.3}$$

where $m(g_1, g_2)$ is a complex number of modulo one (U is said to be a projective representation). Moreover, they showed that the study of projective representations can be reduced to the theory of ordinary unitary representations by enlarging the physical group of symmetries. For example, the rotation group $SO(3)$ has to be replaced by its universal covering group $SU(2)$. The trick of replacing the physical symmetry group G with its universal covering group G^* is so well known in the physics community that the group G^* itself is considered as the true physical symmetry group. However, for the Galilei group the covering group is not enough and one needs even a larger group $\overline{G}$, namely the universal central extension, in order that the unitary (ordinary) representations of $\overline{G}$ exhaust all the possible projective representations of G.

The aim of this book is to present the theory needed to construct the universal central extension from the physical symmetry group in a unified, simple and mathematically coherent way. Most of the results presented are known. However, we hope that our exposition will help the reader to understand the role of the mathematical objects that are introduced in order to take care of the true projective character of the representations in quantum mechanics. Finally, our construction of $\overline{G}$ is very explicit and can be performed by simple linear algebraic tools. This theory is presented in Chap. 3.

Coming back to (0.1), this equality means that we regard *symmetries* as mathematical objects that preserve the transition probability between pure states. The structure of transition probability is only one of the various physically relevant structures associated with a quantum system. Other relevant structures being, for instance, the convex structures of the sets of states and effects, the order structure of effects, and the algebraic structure of observables. Therefore it is natural to define *symmetry* as a bijective map that preserves one of these structures. In Chap. 2 we present several possibilities of modeling a symmetry and we show that they all coincide. Hence one may speak of *symmetries* of a quantum system. The set of all possible symmetries forms a topological group Σ and, given a group G, a *symmetry action* is defined as a continuous map σ from G to Σ such that

$$\sigma_{g_1 g_2} = \sigma_{g_1} \sigma_{g_1} .$$

As an application of these ideas, in Chaps. 4 and 5 we treat in full detail the case of the Galilei group both in $3+1$ and in $2+1$ dimensions. The choice of the Galilei group instead of the Poincaré group is motivated first of all by the fact that the Poincaré group has already been extensively studied in the literature. Secondly, from a mathematical point of view, the Galilei group shows all the pathologies cited above and one needs the full theory of projective representations. We also treat the $2+1$ dimensional case since there is an increasing interest in the surface phenomena both from theoretical and from experimental points of view.

The last chapter of the book is devoted to the study of Galilei invariant wave equations. Within the framework of the first quantisation, the need for wave equations naturally arises if one introduces the interaction of a particle with a (classical) electromagnetic field by means of the minimal coupling principle. To this aim, one has to describe the vector states as functions on the space-time satisfying a differential equation, the wave equation, which is invariant with respect to the universal central extension of the Galilei group. In Chap. 6 we describe how these wave equations can be obtained without using Lagrangian (classical) techniques. In particular, we prove that for a particle of spin j there exists a linear wave equation, like the Dirac equation for the Poincaré group, such that the particle acquires a gyromagnetic internal moment with the gyromagnetic ratio $\frac{1}{j}$.

Since the book is devoted to the application of the abstract theory to the Galilei group, we always assume that the symmetry group G is a connected Lie group. In particular, we do not consider the problem of discrete symmetries. In the Appendix we recall some basic mathematical definitions, facts, and theorems needed in this book. The reader will find them as entries in the Dictionary of Mathematical Notions in the Appendix. The statement of definitions and results are usually not given in their full generality but they are adjusted to our needs.

Contents

1 A Synopsis of Quantum Mechanics

This chapter collects the basic elements of quantum mechanics in the form that is appropriate for an analysis of space-time symmetries. The reader who is familiar with the Hilbert space formulation of quantum mechanics may start directly with Chap. 2 of the book and return here if a need to check our notations and terminology arises.

In quantum mechanics a physical system is represented by means of a complex separable Hilbert space $\mathcal{H}$, with an inner product $\langle \cdot, \cdot \rangle$. The general structure of any experiment – a preparation of a system, followed by a measurement on it – is reflected in the concepts of states and observables, or, states and effects. In their most rudimentary forms states and observables of the system are given, respectively, as unit vectors $\varphi \in \mathcal{H}$ and selfadjoint operators A acting on $\mathcal{H}$. The real number $\langle \varphi, A\varphi \rangle$ is then interpreted as the expectation value of the measurement outcomes of the observable A when measured repeatedly on the system in the same state φ.

The probabilistic content of the 'expectation value postulate' becomes more transparent when one considers the spectral decomposition of A. Indeed, if $A = \int_{\mathbb{R}} x d\Pi^A(x)$ is the spectral decomposition of A, then for any unit vector φ the number $\langle \varphi, A\varphi \rangle$ is just the expectation value of the probability measure $X \mapsto \langle \varphi, \Pi^A(X)\varphi \rangle$, where $\Pi^A(X)$ is the spectral projection of A associated with the Borel subsets X of the real line $\mathbb{R}$. The number $\langle \varphi, \Pi^A(X)\varphi \rangle \in [0, 1]$ is interpreted as the probability that a measurement of A leads to a result in the set X when the system is in the state φ.

Both theoretical and experimental reasons require a slight generalisation of the above framework. First of all, in order to take into account statistical mixtures and to describe states of subsystems of compound systems one also needs density matrices: vector states and density matrices are simply the *states* of the system and are represented by positive trace one operators. Moreover, in order to give a probabilistic interpretation to the theory, the only requirement is that the map $X \mapsto \langle \varphi, \Pi^A(X)\varphi \rangle$ is a probability measure on $\mathbb{R}$. Hence, one may replace the projection operator $\Pi^A(X)$ with a positive operator $E(X)$ such that $E(X)$ is bounded by the identity operator I: such an operator is called an *effect* of the system. An observable is then given as an effect valued measure $X \mapsto E(X)$.

G. Cassinelli, E. De Vito, P.J. Lahti, and A. Levrero, *The Theory of Symmetry Actions in Quantum Mechanics*, Lect. Notes Phys. **654**, pp. 1–6
http://www.springerlink.com/

In this generality, if $\mathrm{tr}[\cdot]$ denotes the trace of a trace class operator, then the real number $\mathrm{tr}[TE] \in [0,1]$ gives the probability for an effect E in a state T.

In the next two sections we shall have a closer look at the basic sets of states and effects emphasizing those structures which lead to natural formulations of symmetry transformations. We end this chapter with a brief remark on the notion of an observable. The material presented here is quite standard. For further information on the basic structures of quantum mechanics the reader may consult, in addition to the classics of von Neumann [36] and Dirac [12], any of her or his favorite books on the subject matter. Most of the results quoted here are presented in a more detailed form, for instance, in the monographs of Beltrametti and Cassinelli [3], Busch et al [8], Davies [11], Holevo [19, 20], Jauch [23], Ludwig [25], or Varadarajan [35].

1.1 The Set S of States and the Set P of Pure States

Let $\mathcal{H}$ be the Hilbert space of the quantum system, with inner product $\langle \cdot, \cdot \rangle$, linear in the second argument. Let $\mathbf{B}$ denote the set of bounded operators on $\mathcal{H}$ and let $\mathbf{B}_1$ be its subset of the trace class operators. We denote by $\mathrm{tr}[T]$ the trace of an element $T \in \mathbf{B}_1$. If A, B are in $\mathbf{B}$, we write $A \leq B$, or $B \geq A$, if $B - A$ is a positive operator.

A state T of the system is an element of $\mathbf{B}_1$ such that T is positive and of trace one. We let $\mathbf{S}$ be the set of all states, that is,

$$\mathbf{S} := \{T \in \mathbf{B}_1 \mid T \geq O, \ \mathrm{tr}[T] = 1\}. \tag{1.1}$$

It is a convex subset of the set $\mathbf{B}_1$. Indeed, if $T_1, T_2 \in \mathbf{S}$ and $0 \leq w \leq 1$, then $wT_1 + (1-w)T_2 \in \mathbf{S}$. In fact, $\mathbf{S}$ is even σ-convex, that is, if $(T_i)_{i=1}^{\infty}$ is a sequence of states and $(w_i)_{i=1}^{\infty}$ is a sequence of numbers such that $0 \leq w_i \leq 1, \sum_{i=1}^{\infty} w_i = 1$, then the series $\sum_{i=1}^{\infty} w_i T_i$ converges in $\mathbf{B}_1$ in the trace norm $\|\cdot\|_1$ to an operator in $\mathbf{S}$; we denote this state as $\sum w_i T_i$.

The convex structure of $\mathbf{S}$ reflects the physical possibility of combining states into new states by mixing them with given weights. If $T = wT_1 + (1-w)T_2$, we say that T is a mixture of the states T_1 and T_2 with the weight w. The convex structure of $\mathbf{S}$ allows one to identify its extreme elements, that is, the elements $T \in \mathbf{S}$ for which the condition $T = wT_1 + (1-w)T_2$, with $T_1, T_2 \in \mathbf{S}, 0 < w < 1$, is fulfilled only for $T = T_1 = T_2$. The extreme states are thus those states which cannot be expressed as mixtures of other states. Such states are often called pure states, a notion which, however, requires further qualification in the presence of the so-called superselection rules. We let $\mathrm{ex}\,(\mathbf{S})$ denote the set of extreme states.

For any $\varphi \in \mathcal{H}$, $\varphi \neq 0$, we let $P[\varphi]$ denote the projection on the one-dimensional subspace $[\varphi] := \{c\varphi \mid c \in \mathbb{C}\}$ generated by φ, that is,

$$P[\varphi]\psi := \frac{\langle \varphi, \psi \rangle}{\langle \varphi, \varphi \rangle}\, \varphi,$$

for all $\psi \in \mathcal{H}$. Let $\mathbf{P}$ denote the set of one-dimensional projections on $\mathcal{H}$. Then for any $P \in \mathbf{P}$, $P = P[\varphi]$ for some nonzero $\varphi \in P(\mathcal{H})$, the range of P.

The set $\mathbf{P}$ is an important subset of $\mathbf{S}$. Indeed, if $T \in \mathbf{S}$, then T, as a compact selfadjoint operator has a decomposition $T = \sum_{i=0}^{\infty} w_i P_i$, where (P_i) is a mutually orthogonal $(P_i P_j = O)$ sequence in $\mathbf{P}$, $w_i \in [0,1], \sum w_i = 1$, with the series converging in the operator norm of $\mathbf{B}$ (since T is compact) but also in the trace norm of $\mathbf{B}_1$ (since T is trace class). The numbers w_i, $w_i \neq 0$, are the nonzero eigenvalues of T, each of them occurring in the decomposition as many times as given by the (finite) dimension of the corresponding eigenspace. On the basis of this result it is straightforward to show that the set of extreme states is equal to the set of one-dimensional projections,

$$\mathrm{ex}\,(\mathbf{S}) = \mathbf{P}. \tag{1.2}$$

For this reason we also call the extreme states the vector states. The above result also shows that the σ-convex hull of $\mathbf{P}$ is the whole set of states,

$$\sigma - \mathrm{co}\,(\mathbf{P}) = \mathbf{S}. \tag{1.3}$$

In other words, vector states exhaust all states in the sense that any state can be expressed as a mixture of at most countably many vector states.

It is a basic feature of quantum mechanics that any two (or more) vector states P_1 and P_2 can also be combined into a new vector state by superposing them. To describe this familiar notion in an appropriate way, let $P_1 \vee P_2$ denote the least upper bound of P_1 and P_2. Then any $P \in \mathbf{P}$ which is contained in $P_1 \vee P_2$, that is, $P \leq P_1 \vee P_2$, is a superposition of P_1 and P_2. On the other hand, any vector state P can be expressed as a superposition of a vector state P_1 and another vector state P_2 exactly when P_1 is not orthogonal to P, $P_1 \not\leq P^{\perp}$, that is, if and only if $\mathrm{tr}\big[PP_1\big] \neq 0$ (we are excluding here the trivial case $P_2 = P$).

As is well-known, the idea of superposition of vector states is most directly expressed using the linear structure of the underlying Hilbert space. Indeed, if $P_1 = P[\varphi_1]$ and $P_2 = P[\varphi_2]$, then the superpositions of P_1 and P_2 are exactly those vector states which are of the form $P = P[c_1\varphi_1 + c_2\varphi_2]$, with $c_1, c_2 \in \mathbb{C}$. If $P = P[\varphi]$ is any vector state and $P_1 = P[\varphi_1]$ is such that $P_1 \not\leq P^{\perp}$, then $\langle \varphi, \varphi_1 \rangle \neq 0$, and P is a superposition of P_1 and, for instance, $P[\varphi - \langle \varphi_1, \varphi \rangle \varphi_1]$.

1.2 The Set E of Effects and the Set D of Projections

Any state $T \in \mathbf{S}$ induces an expectation functional $E \mapsto \mathrm{tr}\big[TE\big]$ on the set $\mathbf{B}$ of bounded operators. The requirement that the numbers $\mathrm{tr}\big[TE\big]$ represent probabilities implies that the operator E is positive and bounded by the unit operator: $O \leq E \leq I$. Such operators are called effects and the number $\mathrm{tr}\big[TE\big]$ is the probability for the effect E in the state T. Let

$$\mathbf{E} := \{E \in \mathbf{B} \,|\, O \leq E \leq I\} \tag{1.4}$$

denote the set of all effects.

As a subset of **B**, **E** is ordered, with O and I as its order bounds. The order on **E** is connected with the basic probabilities of quantum mechanics. Indeed, for any $E, F \in \mathbf{E}$, $E \leq F$ (in the sense that $F - E \geq O$) if and only if $\mathrm{tr}[TE] \leq \mathrm{tr}[TF]$ for all $T \in \mathbf{S}$. The map $\mathbf{E} \ni E \mapsto E^\perp := I - E \in \mathbf{E}$ is a kind of complementation, since it reverses the order (if $E \leq F$, then $F^\perp \leq E^\perp$) and, when applied twice, it yields the identity ($(E^\perp)^\perp = E$). These properties guarantee that the de Morgan laws hold in **E** in the sense that if, for instance, the greatest lower bound $E \wedge F$ of $E, F \in \mathbf{E}$ exists in **E**, then the least upper bound of their complements $E^\perp$ and $F^\perp$ also exists in **E** and $(E \wedge F)^\perp = E^\perp \vee F^\perp$. However, $E \mapsto E^\perp$ is not an orthocomplementation since the greatest lower bound of E and $E^\perp$ need not exist at all, or, even when it does, it need not be the null effect.

The set of projections **D** is an important subset of **E**. For any $E \in \mathbf{E}$, $EE^\perp = E^\perp E$, so that $EE^\perp$ is an effect contained in both E and $E^\perp$. Therefore, the projections can be characterized as those effects E for which the set of lower bounds of E and $E^\perp$, $\mathrm{l.b.}\,\{E, E^\perp\}$, contains only the null effect,

$$\mathbf{D} = \{D \in \mathbf{E} \,|\; \mathrm{l.b.}\,\{D, D^\perp\} = \{O\}\,\}. \tag{1.5}$$

In addition to its order structure, the set **E** of effects is a convex subset of the set of bounded operators **B**: for any $E, F \in \mathbf{E}$ and $0 \leq w \leq 1$, $wE + (1-w)F \in \mathbf{E}$. This structure reflects the physical possibility of combining measurements into new measurements by mixing them. An effect $E \in \mathbf{E}$ is an extreme effect if the condition $E = wE_1 + (1-w)E_2$, with $E_1, E_2 \in \mathbf{E}$, $0 < w < 1$, implies that $E = E_1 = E_2$. Extreme effects arise from pure measurements, that is, measurements which cannot be obtained by mixing some other measurements. By a straightforward application of the spectral theorem one may show that the set of extreme effects $\mathrm{ex}\,(\mathbf{E})$ equals with the set of projections,

$$\mathrm{ex}\,(\mathbf{E}) = \mathbf{D}. \tag{1.6}$$

The algebraic structure of **B** also equips **E** with the structure of a partial algebra. Indeed, for any $E, F \in \mathbf{E}$, their sum $E + F$ is an effect whenever the operator $E + F$ is bounded by the unit operator. Moreover, for each $E \in \mathbf{E}$, there is a unique $E' \in \mathbf{E}$ such that $E + E' = I$. Clearly, $E' = E^\perp$. This structure is closely related to the physical possibility that the effects E and F, for which $E + F \leq I$, can be measured together. The partial sum leads us to define an order on **E**: for any $E, F \in \mathbf{E}$, we write $E \leq F$ exactly when there is a $G \in \mathbf{E}$ such that $E + G = F$. Obviously, the order so defined agrees with the order given by the notion of a positive operator. We observe also that if $D_1, D_2 \in \mathbf{D}$, then $D_1 + D_2$ is an effect if and only if it is a projection, hence **D** itself is endowed with a partial algebra structure by restricting on it the partially defined sum of $\mathbf{E}$. The order defined on $\mathbf{D}$ by this partial sum is obviously the standard one.

There is, however, an important difference between $\mathbf{D}$ and $\mathbf{E}$ as concerns the relation between their structures of partial algebras and ortho-ordered sets. In fact, given $D_1, D_2 \in \mathbf{D}$, one has $D_1 + D_2 \in \mathbf{D}$ if and only if $D_1 \le D_2^\perp$ and, in this case, $D_1 + D_2 = D_1 \vee D_2$. Hence, not only the partial algebra structure of $\mathbf{D}$ determines its order structure, but the converse is also true. This is, however, not the case in the set of effects. In fact, there exist the effects $E, F \in \mathbf{E}$ such that $E \le F^\perp$ and $E + F \in \mathbf{E}$, but $E + F \neq E \vee F$, as would be required if we were to define the partial sum in terms of the order. This is due to the fact that $E \vee F$ need not exist at all. As an example, consider $E = \alpha D_1, F = \beta D_2$, with $0 < \alpha < \beta < 1$, $D_1 \le D_2^\perp$ and $D_1, D_2 \in \mathbf{D}$. Then $\alpha D_1 \le (\beta D_2)^\perp$, $\alpha D_1 + \beta D_2 \in \mathbf{E}$, but $\alpha D_1 \vee \beta D_2$ does not exist.

With respect to the partial sum structure, the projections may again be distinguished as a special subset of effects. Indeed, $\mathbf{D}$ is the set of effects $E \in \mathbf{E}$ for which the set of upper bounds $\text{u.b.}\{E, E'\} = \{I\}$, in the order given by the sum.

The notion of the coexistence of effects is a fundamental concept in quantum mechanics which is introduced to describe effects that can be measured together by measuring a single observable. For any two effects $E, F \in \mathbf{F}$ their coexistence can equivalently be formulated as follows: E and F are in coexistence if and only if there are effects $E_1, F_1, G \in \mathbf{E}$ such that $E = E_1 + G, F = F_1 + G$, and $E_1 + F_1 + G \le I$. When applied to projections $D_1, D_2 \in \mathbf{D} \subset \mathbf{E}$, their coexistence is equivalent to their compatibility, which, in turn, is equivalent to the commutativity of D_1 and D_2.

1.3 Observables

We close this introductory chapter with a short remark on observables. In accordance with the idea that an observable provides a representation of the possible events occurring as outcomes of a measurement, we define an observable as an effect valued measure $\Pi : \mathcal{F} \to \mathbf{E}$ on a σ-algebra $\mathcal{F}$ of subsets of a nonempty set Ω. That is, a function $\Pi : \mathcal{F} \to \mathbf{B}$ is an observable if 1) $\Pi(X) \ge O$ for all $X \in \mathcal{F}$, 2) $\Pi(\Omega) = I$, and 3) $\Pi(\cup X_i) = \sum \Pi(X_i)$ for all disjoint sequences $(X_i) \subset \mathcal{F}$, where the series converges in the weak, or, equivalently in the strong operator topology of $\mathbf{B}$. We recall that an observable $\Pi : \mathcal{F} \to \mathbf{B}$ is projection valued, that is, $\Pi(X) \in \mathbf{D}$ for all $X \in \mathcal{F}$, if and only if it is multiplicative, that is, $\Pi(X \cap Y) = \Pi(X)\Pi(Y)$ for all $X, Y \in \mathcal{F}$. Finally, we note that an observable $\Pi : \mathcal{F} \to \mathbf{B}$ and a state $T \in \mathbf{S}$ define a probability measure

$$p_T^\Pi : \mathcal{F} \to [0,1],\ X \mapsto p_T^\Pi(X) := \operatorname{tr}\big[T\Pi(X)\big] ,$$

which, in the minimal interpretation of quantum mechanics, is the probability distribution of the measurement outcomes of Π in state T in the following sense: the number $p_T^{\Pi}(X)$ is the probability that a measurement of the observable Π on the system in the state T leads to a result in the set X. In accordance with this interpretation, the number $\mathrm{tr}\left[TE\right]$ is the probability for the effect $E \in \mathbf{E}$ in the state $T \in \mathbf{S}$, and, since $\mathbf{P} \subset \mathbf{S}$ and $\mathbf{P} \subset \mathbf{E}$, the number $\mathrm{tr}\left[P_1 P_2\right]$ may also be interpreted as the transition probability between the vector states P_1 and P_2.

2 The Automorphism Group of Quantum Mechanics

The idea of symmetry receives its natural mathematical modelling as a transformation on the set of entities the symmetry refers to. The basic structures of quantum mechanics are coded in the sets of states and effects and in the duality between them. As described in Chap. 1 these sets possess various physically relevant structures which define the corresponding automorphism groups. Any of them could be used to formulate the notion of symmetry in quantum mechanics. The plurality here, however, is deceptive since all these automorphism groups turn out to be isomorphic in a natural way. This chapter is devoted to the study of several such groups and the natural connections between them.

Section 2.1 formulates the relevant automorphisms and investigates their main properties. Section 2.2 states and proofs the fundamental representation theorem, the Wigner theorem, for such automorphisms. Section 2.3 summarizes and completes the study of the isomorphisms of the groups of state and effect automorphisms.

We let $\mathcal{H}$ be the Hilbert space of the system and we use the notations and terminology introduced in Chap. 1.

2.1 Automorphism Groups of Quantum Mechanics

The various structures of the sets of states and effects and the function $(T, E) \mapsto \mathrm{tr}\big[TE\big]$ lead to several natural automorphisms of quantum mechanics. They will be discussed in the following subsections.

2.1.1 State Automorphisms

The set $\mathbf{S}$ of states is a convex set, the convexity structure exhibiting the possibility of combining states into new states by mixing them. This structure leads to the following definition of a state automorphism.

Definition 1. A function $s : \mathbf{S} \to \mathbf{S}$ is a *state automorphism* if
1) s is a bijection,
2) $s(wT_1 + (1-w)T_2) = ws(T_1) + (1-w)s(T_2)$ for all $T_1, T_2 \in \mathbf{S}$, $0 \le w \le 1$.

G. Cassinelli, E. De Vito, P.J. Lahti, and A. Levrero, *The Theory of Symmetry Actions in Quantum Mechanics*, Lect. Notes Phys. **654**, pp. 7–25
http://www.springerlink.com/

Let $\mathrm{Aut}(\mathbf{S})$ denote the set of all state automorphisms. It is straightforward to confirm that $\mathrm{Aut}(\mathbf{S})$ is a group with respect to the composition of functions. The duality $(T,E) \mapsto \mathrm{tr}[TE]$ serves to define a natural topology on $\mathrm{Aut}(\mathbf{S})$. Indeed, any pair of a state T and an effect E defines a function $\mathrm{Aut}(\mathbf{S}) \ni s \mapsto f_{T,E}(s) := \mathrm{tr}[s(T)E] \in [0,1]$, and we endow $\mathrm{Aut}(\mathbf{S})$ with the weakest topology in which all these functions $f_{T,E}$, $T \in \mathbf{S}$, $E \in \mathbf{E}$, are continuous. The following lemma gives some basic properties of state automorphisms.

Lemma 1. *Let $s \in \mathrm{Aut}(\mathbf{S})$.*
1) s is the restriction of a unique trace-norm preserving linear operator on the set $\mathbf{B}_{1,r}$ of the selfadjoint trace class operators on $\mathcal{H}$;
2) $s(\mathbf{P}) \subseteq \mathbf{P}$;
3) if $s(P) = P$ for all $P \in \mathbf{P}$, then s is the identity.

Proof. 1) To extend s to $\mathbf{B}_{1,r} := \{T \in \mathbf{B}_1 \,|\, T^* = T\}$ consider first a $T \in \mathbf{B}_{1,r}^+ := \{T \in \mathbf{B}_{1,r} \,|\, T \geq O\}$, and define

$$\tilde{s}(T) := \|T\|_1 \, s(T/\|T\|_1)$$

for $T \neq O$ and put $\tilde{s}(T) = O$ if $T = O$. Then, for any $\lambda \geq 0$, one gets $\tilde{s}(\lambda T) = \lambda \tilde{s}(T)$, which is the positive homogeneity of $\tilde{s}$. Now let $T_1, T_2 \in \mathbf{B}_{1,r}^+$ and write $T_1 + T_2$ in the form

$$T_1 + T_2 = (\|T_1\|_1 + \|T_2\|_1) \left(\frac{\|T_1\|_1}{\|T_1\|_1 + \|T_2\|_1} \frac{T_1}{\|T_1\|_1} + \frac{\|T_2\|_1}{\|T_1\|_1 + \|T_2\|_1} \frac{T_2}{\|T_2\|_1} \right).$$

The positive homogeneity of $\tilde{s}$ and the convexity of s then yield the additivity of $\tilde{s}$, $\tilde{s}(T_1+T_2) = \tilde{s}(T_1)+\tilde{s}(T_2)$. Consider next a $T \in \mathbf{B}_{1,r}$, write $T = T^+ - T^-$, where $T^{\pm} = \frac{1}{2}(|T| \pm T)$, with $|T| := \sqrt{T^*T}$, and define

$$\hat{s}(T) := \tilde{s}(T^+) - \tilde{s}(T^-).$$

The additivity of $\tilde{s}$ and its homogeneity over non-negative real numbers give the linearity of $\hat{s}$. Also, if $T = T_1 - T_2$ for some other $T_1, T_2 \in \mathbf{B}_{1,r}^+$, then $T^+ + T_2 = T^- + T_1$, so that by the additivity of $\tilde{s}$, $\tilde{s}(T^+) - \tilde{s}(T^-) = \tilde{s}(T_1) - \tilde{s}(T_2)$, which shows that $\hat{s}$ is well defined. By construction, $\hat{s}$ is positive, that is, $\hat{s}(T) \geq O$ for all $T \geq O$. Moreover, it preserves the trace, since

$$\begin{aligned} \mathrm{tr}[\hat{s}(T)] &= \mathrm{tr}\,\left[\|T^+\|_1 \, s(T^+/\|T^+\|_1) - \|T^-\|_1 \, s(T^-/\|T^-\|_1)\right] \\ &= \|T^+\|_1 - \|T^-\|_1 = \mathrm{tr}[T^+] - \mathrm{tr}[T^-] = \mathrm{tr}[T] \end{aligned}$$

for all $T \in \mathbf{B}_{1,r}$. If $f : \mathbf{B}_{1,r} \to \mathbf{B}_{1,r}$ is another positive linear map which extends s, then for any $T \in \mathbf{B}_{1,r}$, $f(T) = f(T^+ - T^-) = f(T^+) - f(T^-) = \|T^+\|_1 f(T^+/\|T^+\|_1) - \|T^-\|_1 f(T^-/\|T^-\|_1) = \|T^+\|_1 s(T^+/\|T^+\|_1) - \|T^-\|_1 s(T^-/\|T^-\|_1) = \hat{s}(T)$, showing that the extension is unique. A direct computation shows, in addition, that $\widehat{s^{-1}}$ is the inverse of $\hat{s}$ so that $\hat{s}$ is a bijection.

It remains to be shown that $\hat{s}$ preserves the trace norm. In fact, for any $T \in \mathbf{B}_{1,r}$, we have

$$\begin{aligned}
\|\hat{s}(T)\|_1 &= \left\|\hat{s}(T^+ - T^-)\right\|_1 = \left\|\hat{s}(T^+) - \hat{s}(T^-)\right\|_1 \\
&\leq \left\|\hat{s}(T^+)\right\|_1 + \left\|\hat{s}(T^-)\right\|_1 = \left\|T^+\right\|_1 + \left\|T^-\right\|_1 \\
&= \mathrm{tr}\left[T^+ + T^-\right] = \mathrm{tr}\left[|T|\right] = \|T\|_1 \,.
\end{aligned}$$

Since the inverse s^{-1} of s has the same properties as s, one also has $\|T\|_1 = \left\|\hat{s}^{-1}(\hat{s}(T))\right\|_1 \leq \|\hat{s}(T)\|_1$, so that $\|\hat{s}(T)\|_1 = \|T\|_1$.

2) Let $P \in \mathbf{P}$ and assume that $s(P) = wT_1 + (1-w)T_2$ for some $0 < w < 1, T_1, T_2 \in \mathbf{S}$. Then $P = ws^{-1}(T_1) + (1-w)s^{-1}(T_2)$, so that $P = s^{-1}(T_1) = s^{-1}(T_2)$ and thus $s(P) = T_1 = T_2$ showing that $s(P) \in \mathbf{P}$.

3) Any $T \in \mathbf{S}$ can be expressed as $T = \sum_i w_i P_i$ for some sequence (w_i) of weights $[0 \leq w_i \leq 1, \sum w_i = 1]$ and for some sequence of elements $(P_i) \subset \mathbf{P}$ with the series converging in the trace norm. By the continuity of s, $s(T) = \sum_i w_i s(P_i)$, which shows that $s(T) = T$ for all $T \in \mathbf{S}$ whenever $s(P) = P$ for all $P \in \mathbf{P}$. □

Example 1. For any unitary operator $U \in \mathbf{U}$ define $s_U(T) := UTU^*$ for all $T \in \mathbf{S}$. Clearly, s_U is a state automorphism. Let $U_1, U_2 \in \mathbf{U}$. Then $s_{U_1} = s_{U_2}$ if and only if $U_1 = zU_2$ for some complex number z of modulus one. Indeed, if $s_{U_1}(T) = s_{U_2}(T)$ for all $T \in \mathbf{S}$, then, in particular, $s_{U_1}(P) = s_{U_2}(P)$ for all $P \in \mathbf{P}$, so that $U_1\varphi = z(\varphi)U_2\varphi$, $z(\varphi) \in \mathbb{T}$, for all $\varphi \in \mathcal{H}$. It remains to be shown that the function $\varphi \mapsto z(\varphi)$ is constant. Let $c \in \mathbb{C}, \varphi \in \mathcal{H}$. Then $U_1(c\varphi) = cU_1\varphi = cz(\varphi)U_2\varphi$ and $U_1(c\varphi) = z(c\varphi)U_2(c\varphi) = cz(c\varphi)U_2\varphi$, so that $z(\varphi) = z(c\varphi)$. Let $\varphi, \psi \in \mathcal{H}$. Then $U_1(\varphi + \psi) = U_1\varphi + U_1\psi = z(\varphi)U_2\varphi + z(\psi)U_2\psi$ as well as $U_1(\varphi+\psi) = z(\varphi+\psi)U_2(\varphi+\psi) = z(\varphi+\psi)U_2\varphi + z(\varphi+\psi)U_2\psi$. Assume that $\varphi \neq c\psi$, that is, φ and ψ are linearly independent. Then $\theta := (\langle\psi,\psi\rangle\varphi - \langle\psi,\varphi\rangle\psi) \,/\, (\langle\psi,\psi\rangle\langle\varphi,\varphi\rangle - \langle\varphi,\psi\rangle\langle\psi,\varphi\rangle)$ is a vector such that $\langle\theta,\varphi\rangle = 1$ and $\langle\theta,\psi\rangle = 0$. Taking the scalar product of the vector $U_1(\varphi+\psi)$ with the vector $U_2\theta$ then yields $z(\varphi) = z(\varphi+\psi)$ for any $\psi \in \mathcal{H}$ that is linearly independent of φ. Therefore, $z(\varphi)$ is constant. Similarly, if $U \in \overline{\mathbf{U}}$ is an antiunitary operator, then s_U, with $s_U(T) := UTU^*$, $T \in \mathbf{S}$, is an element of $\mathrm{Aut}\,(\mathbf{S})$, and two such automorphisms s_{U_1} and s_{U_2} are exactly the same when $U_1 = zU_2$ for some $z \in \mathbb{T}$.

2.1.2 Vector State Automorphisms

The set $\mathbf{P}$ of vector states is a distinguished subset of the set of all states, $\mathbf{P} = \mathrm{ex}\,(\mathbf{S})$. These are the states that cannot be expressed as mixtures of other states. However, they can be superposed into new vector states and any vector state can be expressed as a superposition of some other vector states. We use this structure to define the following notion of an automorphism of vector states.

Definition 2. A function $p : \mathbf{P} \to \mathbf{P}$ is a *superposition automorphism* if
1) p is a bijection,
2) for all $P, P_1, P_2 \in \mathbf{P}$, $P \le P_1 \vee P_2 \iff p(P) \le p(P_1) \vee p(P_2)$,
3) for all $P, P_1 \in \mathbf{P}$, $P_1 \not\le P^\perp \iff p(P_1) \not\le p(P)^\perp$.

Let $\mathrm{Aut}_s(\mathbf{P})$ denote the set of all superposition automorphisms of vector states. It is a group with respect to the composition of functions, and the functions $p \mapsto f_{P,E}(p) := \mathrm{tr}\big[p(P)E\big], P \in \mathbf{P}, E \in \mathbf{E}$, give it a natural initial topology. Given $U \in \mathbf{U} \cup \overline{\mathbf{U}}$ we can define $p_U : \mathbf{P} \to \mathbf{P}$ as $p_U(P) = UPU^*$. Then $p_U \in \mathrm{Aut}_s(\mathbf{P})$ and $p_{U_1} = p_{U_2}$ if and only if $U_1 = zU_2$ for some $z \in \mathbb{T}$.

The notion of transition probability on $\mathbf{P}$ serves to define another natural notion of a vector state automorphism. We simply call it a vector state automorphism.

Definition 3. A function $p : \mathbf{P} \to \mathbf{P}$ is a *vector state automorphism* if
1) p is a bijection,
2) $\mathrm{tr}\big[p(P_1)p(P_2)\big] = \mathrm{tr}\big[P_1P_2\big]$ for all $P_1, P_2 \in \mathbf{P}$.

Let $\mathrm{Aut}\,(\mathbf{P})$ denote the set of all vector state automorphisms. One may again readily check that $\mathrm{Aut}\,(\mathbf{P})$ forms a group with respect to the function composition, $p_U \in \mathrm{Aut}\,(\mathbf{P})$ for each $U \in \mathbf{U} \cup \overline{\mathbf{U}}$ and the basic duality defines a natural topology on $\mathrm{Aut}\,(\mathbf{P})$. This is the initial topology defined by the family of functions f_{P_1,P_2}, $P_1, P_2 \in \mathbf{P}$, where $f_{P_1,P_2}(p) := \mathrm{tr}\big[p(P_1)P_2\big]$.

Condition 3 of Definition 2 is equivalent to the condition that $\mathrm{tr}\big[p(P_1)p(P)\big] = 0$ if and only if $\mathrm{tr}\big[P_1P\big] = 0$. This is a weakening of condition 2 of Definition 3. Let $\mathrm{Aut}_0(\mathbf{P})$ denote the group of the bijective functions $p : \mathbf{P} \to \mathbf{P}$ which satisfy condition 3 of Definition 2, that is, which preserve transition probability zero. Then $\mathrm{Aut}_s(\mathbf{P}) \subseteq \mathrm{Aut}_0(\mathbf{P})$ and $\mathrm{Aut}\,(\mathbf{P}) \subseteq \mathrm{Aut}_0(\mathbf{P})$. We shall see that, if the dimension of the Hilbert space is greater than 2, then these three groups are the same. On the other hand, if $\dim \mathcal{H} = 2$, then $\mathrm{Aut}\,(\mathbf{P}) \subset \mathrm{Aut}_0(\mathbf{P}) = \mathrm{Aut}_s(\mathbf{P})$. The following example exhibits the two dimensional case, whereas we return to confirm the remaining statements in Sect. 2.3.1.

Example 2. Consider the two dimensional Hilbert space $\mathcal{H} = \mathbb{C}^2$. The set $\mathbf{P}$ of one-dimensional projections on $\mathbb{C}^2$ consists exactly of the operators of the form $\frac{1}{2}(I + \boldsymbol{a} \cdot \boldsymbol{\sigma})$, where $\boldsymbol{a} \in \mathbb{R}^3$, $\|\boldsymbol{a}\| = 1$, and $\boldsymbol{\sigma} = (\sigma_1, \sigma_2, \sigma_3)$ are the Pauli matrices. Therefore, any $p : \mathbf{P} \to \mathbf{P}$ is of the form $\frac{1}{2}(I + \boldsymbol{a} \cdot \boldsymbol{\sigma}) \mapsto \frac{1}{2}(I + \boldsymbol{a}' \cdot \boldsymbol{\sigma})$ so that p is bijective if and only if $\boldsymbol{a} \mapsto \boldsymbol{a}' =: f(\boldsymbol{a})$ is a bijection on the unit sphere of $\mathbb{R}^3$. Writing $\boldsymbol{a} = (1, \theta, \phi)$, $\theta \in [0, \pi], \phi \in [0, 2\pi]$ we define a function f such that $f(1, \theta, \phi) = (1, \theta, \phi)$ whenever $\theta \ne \frac{\pi}{2}$ and we write $f(1, \frac{\pi}{2}, \phi) = (1, \frac{\pi}{2}, g(\phi))$, with $g(\phi) = \phi^2/\pi$ for $0 \le \phi \le \pi$ and $g(\phi) = (\phi - \pi)^2/\pi + \pi$ for $\pi \le \phi \le 2\pi$. The function $p : \mathbf{P} \to \mathbf{P}$ defined by f is clearly bijective. Using the fact that $\mathrm{tr}\big[\frac{1}{2}(I + \boldsymbol{a} \cdot \boldsymbol{\sigma})\frac{1}{2}(I + \boldsymbol{b} \cdot \boldsymbol{\sigma})\big] = \frac{1}{2}(1 + \boldsymbol{a} \cdot \boldsymbol{b})$ one immediately observes that p preserves the transition probability zero but not, in general, other transition probabilities. Hence $p \in \mathrm{Aut}_0(\mathbf{P})$, but $p \notin \mathrm{Aut}\,(\mathbf{P})$; this example is essentially due to Uhlhorn [37]. Finally, in the two dimensional case, condition 2 of Definition 2 is trivial, so that now $\mathrm{Aut}_0(\mathbf{P}) = \mathrm{Aut}_s(\mathbf{P})$.

The set $\mathbf{P}$ is a subset of $\mathbf{S}$. One may then ask whether a state automorphism, when restricted to the vector states, defines a vector state automorphism. The following lemma answers this question affirmatively, showing, in fact, that the restriction $s \mapsto s|_{\mathbf{P}}$ defines a group isomorphism $\mathrm{Aut}\,(\mathbf{S}) \to \mathrm{Aut}\,(\mathbf{P})$.

Proposition 1. *The function* $\mathrm{Aut}\,(\mathbf{S}) \ni s \mapsto s|_{\mathbf{P}} \in \mathrm{Aut}\,(\mathbf{P})$ *is a group isomorphism.*

Proof. Let $s \in \mathrm{Aut}\,(\mathbf{S})$. By Lemma 1 its restriction $s|_{\mathbf{P}}$ on $\mathbf{P}$ is well-defined and bijective. Let $\hat{s}$ be the trace-norm preserving linear extension of s on $\mathbf{B}_{1,r}$, and let $P_1, P_2 \in \mathbf{P}$. Then

$$\begin{aligned} 2\sqrt{1 - \mathrm{tr}\big[P_1 P_2\big]} &= \|P_1 - P_2\|_1 = \|\hat{s}\,(P_1 - P_2)\|_1 = \|\hat{s}(P_1) - \hat{s}(P_2)\|_1 \\ &= \|s(P_1) - s(P_2)\|_1 = 2\sqrt{1 - \mathrm{tr}\big[s(P_1)s(P_2)\big]}, \end{aligned}$$

so that $s|_{\mathbf{P}}$ preserves the transition probabilities. The map $s \mapsto s|_{\mathbf{P}}$ is clearly a group homomorphism. Its injectivity follows from the above proved fact that s is the identity whenever $s|_{\mathbf{P}}$ is such. To prove the surjectivity, let $p \in \mathrm{Aut}\,(\mathbf{P})$. Since any $T \in \mathbf{S}$ can be decomposed as $T = \sum w_i P_i$ we may define $s_p(T) := \sum w_i p(P_i)$. If $T = \sum_j w'_j P'_j$ is another decomposition of T, then a direct computation shows that $\sum_j w'_j p(P'_j) = \sum_i w_i p(P_i)$. Thus s_p is well defined. Its convexity, injectivity, and surjectivity can readily be confirmed. Clearly, $s_p|_{\mathbf{P}} = p$, and the proof is complete. □

2.1.3 Effect Automorphisms

The set of effects $\mathbf{E}$ possesses three distinct, physically relevant basic structures, the $\perp$-order structure, the convexity structure, and the partial algebra structure. They all lead to natural notions of effect automorphisms.

Definition 4. A function $e : \mathbf{E} \to \mathbf{E}$ is an *effect* $\perp$*-order automorphism* if
1) e is a bijection,
2) for all $E, F \in \mathbf{E}$, $E \leq F \iff e(E) \leq e(F)$,
3) $e(E^\perp) = e(E)^\perp$ for all $E \in \mathbf{E}$.

Definition 5. A function $e : \mathbf{E} \to \mathbf{E}$ is an *effect sum automorphism* if
1) e is a bijection,
2) for all $E, F \in \mathbf{E}$, $E + F \in \mathbf{E} \iff e(E) + e(F) \in \mathbf{E}$,
3) $e(E + F) = e(E) + e(F)$ whenever $E + F \in \mathbf{E}$.

Definition 6. A function $e : \mathbf{E} \to \mathbf{E}$ is an *effect convex automorphism* if
1) e is a bijection,
2) $e(wE + (1 - w)F) = we(E) + (1 - w)e(F)$ for all $E, F \in \mathbf{E}$, $0 \leq w \leq 1$.

Let $\mathrm{Aut}_o(\mathbf{E})$, $\mathrm{Aut}_s(\mathbf{E})$, and $\mathrm{Aut}_c(\mathbf{E})$ denote the sets of all effect $\perp$-order, sum, and convex automorphisms, respectively. They all form groups and the functions $f_{T,E} : e \mapsto \mathrm{tr}\big[Te(E)\big]$, $T \in \mathbf{S}, E \in \mathbf{E}$, equip them with natural initial topologies. Clearly, the functions e_U, $U \in \mathbf{U} \cup \overline{\mathbf{U}}$, defined as $e_U(E) = UEU^*$, $E \in \mathbf{E}$, belong to any of these groups. Apart from their apparent difference, the sum and convex automorphisms of effects are identical.

Proposition 2. *The groups* $\mathrm{Aut}_s(\mathbf{E})$ *and* $\mathrm{Aut}_c(\mathbf{E})$ *are the same.*

Proof. Analogously with the extension of $s \in \mathrm{Aut}\,(\mathbf{S})$ to $\hat{s} : \mathbf{B}_{1,r} \to \mathbf{B}_{1,r}$ given in Lemma 1, any sum automorphism $e \in \mathrm{Aut}_s(\mathbf{E})$ can uniquely be extended to a positive bijective linear map on $\mathbf{B}_r$, so that its restriction to $\mathbf{E}$ is, in particular, a convex automorphism. Hence $\mathrm{Aut}_s(\mathbf{E}) \subseteq \mathrm{Aut}_c(\mathbf{E})$. Similarly, any convex automorphism $e \in \mathrm{Aut}_c(\mathbf{E})$ extends uniquely to a positive bijective linear map on $\mathbf{B}_r$, and its restriction to $\mathbf{E}$ is also a sum automorphism, $\mathrm{Aut}_c(\mathbf{E}) \subseteq \mathrm{Aut}_s(\mathbf{E})$. □

Proposition 3. $\mathrm{Aut}_s(\mathbf{E})$ *is a subgroup of* $\mathrm{Aut}_o(\mathbf{E})$.

Proof. Let $e \in \mathrm{Aut}_s(\mathbf{E})$. If $E \leq F$ then $F = (F - E) + E$, with $F - E \in \mathbf{E}$, and thus $e(F) = e(F - E) + e(E)$, so that $e(E) \leq e(F)$. Since e^{-1} shares the properties of e, the converse is also true, that is, if $e(E) \leq e(F)$, then $E \leq F$. The bijectivity of e and the fact that $O = \inf \mathbf{E}$ and $I = \sup \mathbf{E}$ imply that $e(O) = O$ and $e(I) = I$. Since $I = e(I) = e(E + E^\perp) = e(E) + e(E^\perp)$, one also has $e(E)^\perp = e(E^\perp)$. □

Remark 1. An effect $\perp$-order automorphism preserves the orthogonality of effects, that is, it has the property 2) of Definition 5. On the other hand, if $e : \mathbf{E} \to \mathbf{E}$ is a bijection such that for any $E, F \in \mathbf{E}$, $E + F \in \mathbf{E}$ if an only if $e(E) + e(F) \in \mathbf{E}$, then e also preserves the order in both directions. Moreover, since for any $E \in \mathbf{E}$, $E^\perp = \sup\{F \in \mathbf{E} \,|\, E + F \leq I\}$, one gets that $e(E^\perp) = e(E)^\perp$, that is, e is a $\perp$-order automorphism.

Remark 2. The notion of coexistence of effects is a fundamental property of effects. Therefore, one could introduce the corresponding automorphism as a bijection $e : \mathbf{E} \to \mathbf{E}$ satisfying the following condition: for any $E, F \in \mathbf{E}$, E and F are coexistent if and only if $e(E)$ and $e(F)$ are coexistent. The map e for which $e(O) = I, e(I) = O$, and $e(E) = E$ otherwise, is an example of such a transformation, showing that coexistence preserving transformation need not preserve the order, and thus does not lead to a useful characterization. However, when combined with an effect order automorphism, that is, property 2) of Definition 4, the preservation of coexistence in the above sense suffices to determine the structure of such automorphisms for $\dim(\mathcal{H}) \geq 3$ [30].

We proceed to show that an effect sum automorphism defines a unique state automorphism. For this the following two lemmas are needed, the first one being a direct consequence of the previous proposition and the result concerning the limits of increasing bounded nets of selfadjoint operators.

Lemma 2. *Let* $e \in \mathrm{Aut}_s(\mathbf{E})$. *Then*
1) if $(E_i)_{i\in I}$ *is any family of elements of* $\mathbf{E}$ *such that* $\sup_{i\in I} E_i \in \mathbf{E}$ *and* $\sup_{i\in I} e(E_i) \in \mathbf{E}$, *then* $\sup_{i\in I} e(E_i) = e\left(\sup_{i\in I} E_i\right)$;
2) if $(E_i)_{i\in I}$ *is an increasing net of elements of* $\mathbf{E}$, *then* $\sup_{i\in I} E_i \in \mathbf{E}$ *and* $\sup_{i\in I} e(E_i) \in \mathbf{E}$, *and* $\sup_{i\in I} e(E_i) = e\left(\sup_{i\in I} E_i\right)$.

Lemma 3. *Let* $m : \mathbf{E} \to [0,1]$ *be a function with the following properties:*
1) if $E + F \leq I$, *then* $m(E+F) = m(E) + m(F)$,
2) if $(E_i)_{i\in I}$ *is an increasing net in* $\mathbf{E}$, *then* $m\left(\sup_{i\in I} E_i\right) = \sup_{i\in I} m(E_i)$.
There is a unique $T \in \mathbf{B}_{1,r}^{+}$ *such that for all* $E \in \mathbf{E}$, $m(E) = \mathrm{tr}\big[TE\big]$.

Proof. We notice first that $m(E) = m(E+O) = m(E) + m(O)$, so that $m(O) = 0$. We prove next that for all $E \in \mathbf{E}$ and $0 < \lambda < 1$,

$$m(\lambda E) = \lambda m(E).$$

If λ is rational this follows from the additivity of m. Let $0 < \lambda < 1$ and let (r_n) be an increasing sequence of positive rationals converging to λ. Then

$$\sup_n (r_n E) = \lambda E$$

and this implies that

$$m(\lambda E) = m\left(\sup_n \{r_n E\}\right) = \sup_n m(r_n E) = \sup_n (r_n m(E)) = \lambda m(E).$$

The (unique) extension of m to a positive linear map $\hat{m} : \mathbf{B}_r \to \mathbb{R}$ is again straightforward.

The map $\hat{m}$ is normal. Indeed, if $(A_i)_{i\in I}$ is an increasing norm bounded positive net in $\mathbf{B}_r$, then, letting $c = \sup_i \|A_i\|$, $(A_i/c)_{i\in I}$ is an increasing net in $\mathbf{E}$ and we have

$$\hat{m}\left(\sup_i A_i\right) = c\hat{m}\left(\sup_i \frac{A_i}{c}\right) = c\sup_i m\left(\frac{A_i}{c}\right) = \sup_i \hat{m}(A_i).$$

Hence $\hat{m}$ is a linear positive normal function on $\mathbf{B}_r$. It is well known that such an $\hat{m}$ defines a unique positive trace class operator T such that $\hat{m}(A) = \mathrm{tr}\big[TA\big]$ for all $A \in \mathbf{B}_r$, see, for instance [11, Lemma 6.1, Chap. 1]. Since $\hat{m}$ is uniquely defined by its restriction m on $\mathbf{E}$ the proof is complete. □

Proposition 4. *Let* $e \in \mathrm{Aut}_s(\mathbf{E})$. *There is a unique* $s_e \in \mathrm{Aut}(\mathbf{S})$ *such that* $s_e(P) = e(P)$ *for all* $P \in \mathbf{P}$. *Moreover, the correspondence* $\mathrm{Aut}_s(\mathbf{E}) \ni e \mapsto s_e \in \mathrm{Aut}(\mathbf{S})$ *is an injective group homomorphism.*

Proof. Let $e \in \mathrm{Aut}_s(\mathbf{E})$. For all $T \in \mathbf{S}$ define the map from $\mathbf{E}$ to $[0,1]$ by $E \mapsto \mathrm{tr}\big[Te^{-1}(E)\big]$. By the above two lemmas there is a unique positive trace class operator T' such that $\mathrm{tr}\big[Te^{-1}(E)\big] = \mathrm{tr}\big[T'E\big]$ for all $E \in \mathbf{E}$. Taking $E = I$ we have $\mathrm{tr}\big[T'\big] = 1$, hence $T' \in \mathbf{S}$. We define s_e from $\mathbf{S}$ to $\mathbf{S}$ as $s_e(T) := T'$ so that $\mathrm{tr}\big[s_e(T)E\big] = \mathrm{tr}\big[Te^{-1}(E)\big]$, for all $E \in \mathbf{E}$. Using this formula it is straightforward to prove that $s_e \in \mathrm{Aut}\,(\mathbf{S})$ and that $e \mapsto s_e$ is a group homomorphism. Moreover, suppose that $s_e(T) = T$ for all $T \in \mathbf{S}$, then $\mathrm{tr}\big[T(E - e^{-1}(E))\big] = 0, \quad E \in \mathbf{E}$, for all $T \in \mathbf{S}$. Hence $E = e^{-1}(E)$ for all $E \in \mathbf{E}$, that is, e is the identity. This shows the injectivity of the map $e \mapsto s_e$ and ends the proof. □

2.1.4 Automorphisms on D

The set $\mathbf{D}$ of projections is a subset of $\mathbf{E}$. In fact, $\mathbf{D} = \mathrm{ex}\,(\mathbf{E})$. As discussed in Chap. 1, the $\perp$-order structure and the partial algebra structure coincide on $\mathbf{D}$. Consequently, Definitions 4 and 5 when applied to $\mathbf{D}$ are the same, and we choose to consider the following notion of an automorphism on $\mathbf{D}$.

Definition 7. *A function $d : \mathbf{D} \to \mathbf{D}$ is a $\mathbf{D}$-automorphism if*
1) d is a bijection,
2) for all $D_1, D_2 \in \mathbf{D}$, $D_1 \le D_2 \iff d(D_1) \le d(D_2)$,
3) $d(D^\perp) = d(D)^\perp$ for all $D \in \mathbf{D}$.

The set $\mathrm{Aut}\,(\mathbf{D})$ of all $\mathbf{D}$-automorphisms is a group with respect to the composition of functions and it is a topological space with respect to the initial topology given by the functions $f_{T,D} : d \mapsto \mathrm{tr}\big[Td(D)\big]$, $T \in \mathbf{S}$, $D \in \mathbf{D}$. Again, the functions d_U, $U \in \mathbf{U} \cup \overline{\mathbf{U}}$, defined as $d_U(D) = UDU^*$, are elements of $\mathrm{Aut}\,(\mathbf{D})$.

Since $\mathbf{D} \subset \mathbf{E}$ one may consider the restriction of an $e \in \mathrm{Aut}_o(\mathbf{E})$ on $\mathbf{D}$. One gets:

Proposition 5. *The function* $\mathrm{Aut}_o(\mathbf{E}) \ni e \mapsto e|_{\mathbf{D}} \in \mathrm{Aut}\,(\mathbf{D})$ *is a group homomorphism.*

Proof. Let $e \in \mathrm{Aut}_o(\mathbf{E})$. Then for any $E, F, G \in \mathbf{E}$, G is a lower bound of E and F if and only if $e(G)$ is a lower bound of $e(E)$ and $e(F)$. Since $\mathbf{D}$ consists exactly of those effects $E \in \mathbf{E}$ for which O is the only lower bound of E and $E^\perp$ one thus has $e(\mathbf{D}) \subseteq \mathbf{D}$. Clearly, $(e_1 \circ e_2)|_{\mathbf{D}} = e_1|_{\mathbf{D}} \circ e_2|_{\mathbf{D}}$ and $e^{-1}|_{\mathbf{D}} = (e|_{\mathbf{D}})^{-1}$. □

The homomorphism of the above lemma is, in fact, injective whenever the dimension of the Hilbert space is, at least, two. We shall prove this result, which is due to Ludwig [25, Theorem 5.21, p. 226], using the following characterization of effects [18]:

Lemma 4. *For any $E \in \mathbf{E}$,*

$$E = \vee_{P \in \mathbf{P}} (E \wedge P) = \vee_{P \in \mathbf{P}} \lambda(E,P)P,$$

where

$$\lambda(E,P) := \sup\{\lambda \in [0,1] \,|\, \lambda P \leq E\}. \tag{2.1}$$

In fact, $\lambda(E,P) = \max\{\lambda \in [0,1] \,|\, \lambda P \leq E\}$, *and if* $\varphi \in \mathcal{H}, \|\varphi\| = 1$, *is such that* $P\varphi = \varphi$, *then* $\lambda(E,P) = \left\|E^{-1/2}\varphi\right\|^{-2}$, *whenever* $\varphi \in \mathrm{ran}(E^{1/2})$, *whereas* $\lambda(E,P) = 0$, *otherwise.*

Proposition 6. *If* $\dim(\mathcal{H}) \geq 2$, *then the function* $\mathrm{Aut}_o(\mathbf{E}) \ni e \mapsto e|_{\mathbf{D}} \in \mathrm{Aut}\,(\mathbf{D})$ *is injective.*

Proof. It suffices to show that if $e \in \mathrm{Aut}(\mathbf{E})$ is such that $e(D) = D$, for all $D \in \mathbf{D}$, then e is the identity function. Therefore, assume that $e(D) = D$, for all $D \in \mathbf{D}$. Then, in particular, $e(P) = P$, for all $P \in \mathbf{P}$. Thus, for any $\gamma \in [0,1], P \in \mathbf{P}$, $e(\gamma P) \leq e(P) = P$, so that

$$e(\gamma P) = \tau(\gamma, P)P \tag{2.2}$$

for some $\tau(\gamma, P) \in [0,1]$. The proof now consists of showing that, for any $\gamma \in [0,1]$ and for any $P \in \mathbf{P}$, $\tau(\gamma, P) = \gamma$. If this is the case, then, for any $E \in \mathbf{E}$,

$$\begin{aligned} e(E) &= \vee_{P \in \mathbf{P}} e(\lambda(E,P)P) \\ &= \vee_{P \in \mathbf{P}} \tau(\lambda(E,P),P)P \\ &= \vee_{P \in \mathbf{P}} \lambda(E,P)P \\ &= E \end{aligned}$$

and we are through. We proceed in three steps.

Step 1. Let $E \in \mathbf{E}$. From (2.1) we obtain that

$$e(E) = \vee_{P \in \mathbf{P}} e(\lambda(E,P)P) = \vee_{P \in \mathbf{P}} \tau(\lambda(E,P),P)P$$

and also that

$$e(E) = \vee_{P \in \mathbf{P}} \lambda(e(E),P)P.$$

Taking the meet of both expressions with any 1 dimensional projection we see that

$$\tau(\lambda(E,P),P) = \lambda(e(E),P) \tag{2.3}$$

for any $E \in \mathbf{E}, P \in \mathbf{P}$.

Step 2. We show next that the function τ does not depend on P, that is,

$$\tau(\gamma, P) = \tau(\gamma) \tag{2.4}$$

for each $\gamma \in [0, 1]$, $P \in \mathbf{P}$. Clearly $\tau(0, P) = 0$ and $\tau(1, P) = 1$ for all $P \in \mathbf{P}$. Thus, consider a fixed $0 < \gamma < 1$ and let $P, Q \in \mathbf{P}$ be such that $QP \neq O$. Define

$$\mu = \frac{1 - \gamma}{1 - \gamma(1 - \mathrm{tr}[PQ])}. \tag{2.5}$$

Observe that $1 - \gamma \leq \mu < 1$ and define $E := I - \mu Q$. Then $\mathrm{ran}(E^{1/2}) = \mathcal{H}$, so that, by Lemma 4,

$$\lambda(E, P) = \frac{1}{\mathrm{tr}[E^{-1}P]} = \frac{\mu - 1}{\mu(1 - \mathrm{tr}[QP]) - 1} = \gamma.$$

Hence, due to (2.3),

$$\tau(\gamma, P) = \lambda(e(E), P). \tag{2.6}$$

On the other hand $e(E) = I - \tau(\mu, Q)Q$ and again we have $\mathrm{ran}(e(E)^{1/2}) = \mathcal{H}$, so that,

$$\lambda(e(E), P) = \frac{1}{\mathrm{tr}[e(E)^{-1}P]} = \frac{\tau(\mu, Q) - 1}{\tau(\mu, Q)(1 - \mathrm{tr}[QP]) - 1}. \tag{2.7}$$

Comparing (2.6) and (2.7) we have

$$\tau(\gamma, P) = \frac{\tau(\mu, Q) - 1}{\tau(\mu, Q)(1 - \mathrm{tr}[QP]) - 1}.$$

From (2.5) we get

$$(1 - \mathrm{tr}[PQ]) = \frac{\mu + \gamma - 1}{\mu\gamma},$$

hence

$$\tau(\gamma, P) = \frac{\gamma\mu[\tau(\mu, Q) - 1]}{\tau(\mu, Q)(\mu + \gamma - 1) - \gamma\mu}. \tag{2.8}$$

We then see that $\tau(\gamma, P)$ fulfills (2.8), where Q is any 1-dimensional projection such that $\mathrm{tr}[PQ] \neq 0$ and μ is defined by (2.5). On the other hand, μ depends only on γ and $\mathrm{tr}[PQ]$. Given $P_1, P_2 \in \mathbf{P}$, one can find $Q \in \mathbf{P}$ such that $\mathrm{tr}[P_1 Q] = \mathrm{tr}[P_2 Q] \neq 0$ so that (2.8) implies $\tau(\gamma, P_1) = \tau(\gamma, P_2)$ and this proves that τ does not depend on P. Equation (2.4) is thus established.

Step 3. Now suppose that $\dim \mathcal{H} \geq 2$. It is clear from (2.5) that if $1 - \gamma \leq \alpha < 1$, we can choose $P, Q \in \mathbf{P}$ such that $\mu = \alpha$. Hence (2.8) gives

$$\tau(\gamma) = \frac{\gamma\alpha[\tau(\alpha) - 1]}{\tau(\alpha)(\alpha + \gamma - 1) - \gamma\alpha} \tag{2.9}$$

for all α such that $1-\gamma \leq \alpha < 1$. Choosing $\alpha = 1-\gamma$ in (2.9), since $\gamma \in (0,1)$ is arbitrary, we obtain

$$\tau(\gamma) = 1 - \tau(1-\gamma) \tag{2.10}$$

for any $\gamma \in (0,1)$. Observe now that (2.9) can be rewritten as

$$\tau(\gamma) = \frac{a(\alpha)\gamma}{1+\gamma(a(\alpha)-1)} \qquad 1-\gamma \leq \alpha < 1, \tag{2.11}$$

with $a(\alpha) = \frac{\alpha}{\alpha-1}\frac{\tau(\alpha)-1}{\tau(\alpha)}$. We then obtain from (2.11) that

$$a(\alpha) = \frac{1-\gamma}{\gamma}\frac{\tau(\gamma)}{1-\tau(\gamma)} \qquad 1-\gamma \leq \alpha < 1,$$

from which we conclude that $a(\alpha)$ is a constant. By comparison with (2.10) we see that in fact $a(\alpha) = 1$, so that $\tau(\gamma) = \gamma$ for all $\gamma \in [0,1]$. This concludes the proof. □

Proposition 7. *Let* $p \in \mathrm{Aut}_0(\mathbf{P})$. *There is a unique* $d_p \in \mathrm{Aut}\,(\mathbf{D})$ *such that* $d_p(P) = p(P)$ *for all* $P \in \mathbf{P}$. *Moreover, the map* $\mathrm{Aut}_0(\mathbf{P}) \ni p \mapsto d_p \in \mathrm{Aut}\,(\mathbf{D})$ *is a group isomorphism.*

Proof. In this proof we identify the projection lattice $\mathbf{D}$ with the lattice $\mathbf{M}$ of all closed subspaces of $\mathcal{H}$. Let $p \in \mathrm{Aut}_0(\mathbf{P})$. For all $M \subset \mathcal{H}$, $M \neq \{0\}$, let

$$d_p(M) = \{\psi \in p([\phi]) \ : \phi \in M, \ \phi \neq 0\},$$

and put $d_p(\{0\}) = \{0\}$. We observe that $d_{p^{-1}}(d_p(M)) = \{\Phi \in p^{-1}([\psi]) \ : \psi \in d_p([\phi]), \ \phi \in M, \phi \neq 0\} = \{\Phi \in p^{-1}(p[\phi]) \ : \phi \in M, \ \phi \neq 0\} = \mathbb{C}M$. In the same way we have that $d_p(d_{p^{-1}}(M)) = \mathbb{C}M$. Now let $M \in \mathbf{M}$. We then have $d_p(M^\perp) = d_p(M)^\perp$. In fact, if $\phi \in M$ and $\psi \in M^\perp$ are nonzero vectors, then $p(P[\phi]) \perp p([\psi])$. Hence $d_p(M) \perp d_p(M^\perp)$, $d_p(M^\perp) \subset d_p(M)^\perp$ and, since $M = d_{p^{-1}}(d_p(M))$, one concludes that $d_p(M^\perp) = d_p(M)^\perp$. Moreover, since M is a closed subspace, $d_p(M) = d_p((M^\perp)^\perp) = (d_p(M^\perp))^\perp$, proving that $d_p(M)$ is a closed subspace. We denote by d_p the map from $\mathbf{M}$ to $\mathbf{M}$ sending M to $d_p(M)$. Obviously d_p is bijective and preserves the order and the orthogonality, that is, $d_p \in \mathrm{Aut}\,(\mathbf{D})$. Finally, by construction, $d_p(P) = p(P)$ for all $P \in \mathbf{P}$. A standard calculation shows that the map $p \mapsto d_p$ is a group homomorphism. Its injectivity is obvious. Finally, if $d \in \mathrm{Aut}\,(\mathbf{D})$, then obviously $d(\mathbf{P}) = \mathbf{P}$ and d preserves orthogonality, so that $d|_{\mathbf{P}} \in \mathrm{Aut}_0(\mathbf{P})$ and this shows surjectivity. The proof is now complete. □

Proposition 8. *Let* $\dim(\mathcal{H}) \geq 3$. *Given* $d \in \mathrm{Aut}\,(\mathbf{D})$ *there is a unique* $s_d \in \mathrm{Aut}\,(\mathbf{S})$ *such that* $s_d(P) = d(P)$ *for all* $P \in \mathbf{P}$. *Moreover, the map* $\mathrm{Aut}\,(\mathbf{D}) \ni d \mapsto s_d \in \mathrm{Aut}\,(\mathbf{S})$ *is an injective group homomorphism.*

Proof. Let $d \in \mathrm{Aut}\,(\mathbf{D})$. Since d is a lattice orthoisomorphism on $\mathbf{D}$ the mapping $\mathbf{D} \ni D \mapsto \mathrm{tr}\big[Td^{-1}(D)\big] \in [0,1]$ is a generalized probability measure on $\mathbf{D}$ for all $T \in \mathbf{S}$. According to a theorem of Gleason [17] (which holds if the dimension of $\mathcal{H}$ is greater than 2) there is a unique $T' \in \mathbf{S}$ such that $\mathrm{tr}\big[T'D\big] = \mathrm{tr}\big[Td^{-1}(D)\big]$ for all $D \in \mathbf{D}$. The induced function $T \mapsto T' =: s_d(T)$ is one-to-one onto and it preserves the convex structure of $\mathbf{S}$, that is, $s_d \in \mathrm{Aut}\,(\mathbf{S})$. Clearly the map $d \mapsto s_d$ is a group homomorphism. We show now that $s_d(P) = d(P)$ for all $P \in \mathbf{P}$. It is sufficient to prove that $\mathrm{tr}\big[s_d(P_1)P_2\big] = \mathrm{tr}\big[d(P_1)P_2\big]$, $P_1, P_2 \in \mathbf{P}$. Since s_d, restricted to $\mathbf{P}$, is a $\mathbf{P}$-automorphism we have $\mathrm{tr}\big[s_d(P_1)P_2\big] = \mathrm{tr}\big[P_1 s_d^{-1}(P_2)\big] = \mathrm{tr}\big[P_1 s_{d^{-1}}(P_2)\big] = \mathrm{tr}\big[d(P_1)P_2\big]$. Suppose now that $s_d(T) = T$ for all $T \in \mathbf{S}$. Then $d(P) = P$ for all $P \in \mathbf{P}$. Hence, $d(D) = d(\vee_{P \leq D} P) = \vee_{P \leq D} d(P) = \vee_{P \leq D} P = D$ for any $D \in \mathbf{D}$, which shows that d is the identity, and the map $d \mapsto s_d$ is injective. □

To close this subsection consider an effect $\perp$-order automorphism $e \in \mathrm{Aut}_o(\mathbf{E})$ and let $E = \vee_{\mathbf{P}} \lambda(E,P)P$ be the decomposition of $E \in \mathbf{E}$ given in Lemma 4. Then

$$e(E) = e\left(\vee_{\mathbf{P}} \lambda(E,P)P\right) = \vee_{\mathbf{P}} e(\lambda(E,P)P) = \vee_{\mathbf{P}} \lambda\left(e(\lambda(E,P)P), e(P)\right) e(P).$$

Using arguments similar to those applied in the proof of Proposition 6, Mólnar and Páles showed [29] that $\lambda\left(e(\lambda(E,P)P), e(P)\right) = \lambda(E,P)$ and that $\mathrm{tr}\big[P_1P_2\big] = \mathrm{tr}\big[e(P_1)e(P_2)\big]$ for any two $P_1, P_2 \in \mathbf{P}$. We formulate these results in the form of a lemma.

Lemma 5. *Assume that* $\dim(\mathcal{H}) \geq 2$. *The restriction of any* $e \in \mathrm{Aut}_o(\mathbf{E})$ *in* $\mathbf{P}$ *is a vector state automorphism. Moreover, for any* $E \in \mathbf{E}$, $e(E) = \vee_{\mathbf{P}} \lambda(E,P) e(P)$.

2.1.5 Automorphisms of $\mathcal{H}$

With a slight abuse of language, the automorphisms of the Hilbert space $\mathcal{H}$ are either the bijective linear maps $U : \mathcal{H} \to \mathcal{H}$ that preserve the inner product, that is, $\langle U\varphi, U\psi\rangle = \langle\varphi,\psi\rangle$ for all $\varphi,\psi \in \mathcal{H}$, or the bijective antilinear maps $U : \mathcal{H} \to \mathcal{H}$ that reverse the inner product, that is, $\langle U\varphi, U\psi\rangle = \langle\psi,\varphi\rangle$ for all $\varphi,\psi \in \mathcal{H}$. They are exactly the unitary and antiunitary operators on $\mathcal{H}$. The set $\mathrm{Aut}\,(\mathcal{H}) = \mathbf{U} \cup \overline{\mathbf{U}}$ as well as the quotient space $\Sigma = \mathrm{Aut}\,(\mathcal{H})/\mathbf{T}$, where $\mathbf{T} = \{zI \,|\, z \in \mathbb{T}\}$, are topological groups with the properties described in Dictionary A.1 and in Appendix A.2.

Let $\sigma \in \Sigma$ and $U \in \sigma$. Define the function $g_\sigma : \mathbf{B}_r \to \mathbf{B}_r$, by $g_\sigma(A) := UAU^*$. Applying the arguments of Example 1, one observes that g_σ is well defined and $g_{\sigma_1} = g_{\sigma_2}$ only if $\sigma_1 = \sigma_2$. Moreover, when restricted to any of the sets $\mathbf{P}$, $\mathbf{S}$, $\mathbf{D}$, and $\mathbf{E}$, endowed with any of the relevant structures, g_σ defines a corresponding automorphism. We thus conclude with the following proposition.

Proposition 9. *The map $\sigma \mapsto g_\sigma$ defines an injective group homomorphism of Σ to* $\mathrm{Aut}\,(\mathbf{S})$, $\mathrm{Aut}_s(\mathbf{P})$, $\mathrm{Aut}\,(\mathbf{P})$, $\mathrm{Aut}_0(\mathbf{P})$, $\mathrm{Aut}_o(\mathbf{E})$, $\mathrm{Aut}_s(\mathbf{E})$, *and to* $\mathrm{Aut}\,(\mathbf{D})$.

From now on we denote the restrictions of the functions g_σ to the sets $\mathbf{P}$, $\mathbf{S}$, $\mathbf{D}$, and $\mathbf{E}$, respectively, as p_σ, s_σ, d_σ, and e_σ.

2.2 The Wigner Theorem

This section contains the Wigner theorem and two corollaries to it. The proof presented here, originally published in [9], is a modification of Bargmann's proof [2] of Wigner's classic result [39].

2.2.1 The Theorem

Theorem 1. *Let $p \in \mathrm{Aut}\,(\mathbf{P})$. There is a $U \in \mathbf{U} \cup \overline{\mathbf{U}}$ such that $p = p_U$, that is, $p(P) = UPU^*$ for all $P \in \mathbf{P}$. U is unique up to a phase factor.*

Proof. Fix $p \in \mathrm{Aut}\,(\mathbf{P})$. Let $\omega \in \mathcal{H}, \omega \neq 0$, be a fixed vector and define

$$\mathcal{O}_\omega := \big\{\varphi \in \mathcal{H} \mid \langle \omega, \varphi\rangle > 0\big\}.$$

We observe that $\mathcal{O}_\omega$ is a cone, that is, $\mathcal{O}_\omega + \mathcal{O}_\omega \subset \mathcal{O}_\omega$ and $\lambda\mathcal{O}_\omega \subset \mathcal{O}_\omega, \lambda > 0$. Let ω' be a vector in the range of the projection $p(P[\omega])$ such that $\|\omega'\| = \|\omega\|$ and define the cone $\mathcal{O}_{\omega'}$. The proof of the theorem will now be split into five parts.
Part 1. We show that there is a function

$$T_\omega : \mathcal{O}_\omega \to \mathcal{O}_{\omega'}$$

such that for all $\varphi, \varphi_1, \varphi_2 \in \mathcal{O}_\omega, \lambda > 0$,

$$\|T_\omega\varphi\| = \|\varphi\|\,, \tag{2.12}$$
$$T_\omega(\lambda\varphi) = \lambda T_\omega\varphi, \tag{2.13}$$
$$T_\omega(\varphi_1 + \varphi_2) = T_\omega\varphi_1 + T_\omega\varphi_2, \tag{2.14}$$
$$P[T]_\omega\varphi = p(P[\varphi]). \tag{2.15}$$

To define T_ω we observe first that for any vector $\varphi \in \mathcal{O}_\omega$, there is a unique vector $\psi \in \mathcal{O}_{\omega'}$, $\|\psi\| = \|\varphi\|$, such that $p(P[\varphi]) = P[\psi]$. We denote $\psi = T_\omega\varphi$. This defines a function $T_\omega : \mathcal{O}_\omega \to \mathcal{O}_{\omega'}$. Observe that $T_\omega\omega = \omega'$. By definition, T_ω is norm preserving, positively homogeneous, and $p(P[\varphi]) = P[T_\omega\varphi]$. Also for any $\varphi_1, \varphi_2 \in \mathcal{O}_\omega$,

$$|\langle T_\omega\varphi_1, T_\omega\varphi_2\rangle| = |\langle\varphi_1, \varphi_2\rangle|. \tag{2.16}$$

We prove next the additivity of T_ω. Let $\varphi_1, \varphi_2 \in \mathcal{O}_\omega$. By the definition of $\mathcal{O}_\omega$, φ_1 and φ_2 are linearly dependent (over $\mathbb{C}$) if and only if $\varphi_1 = \lambda\varphi_2$ for some $\lambda > 0$. If $\varphi_1 = \lambda\varphi_2$ then $T_\omega(\varphi_1 + \varphi_2) = T_\omega\big((\lambda+1)\varphi_2\big) = (\lambda + 1)T_\omega\varphi_2 = \lambda T_\omega\varphi_2 + T\varphi_2 = T_\omega\varphi_1 + T_\omega\varphi_2$. Assume now that φ_1, φ_2 are linearly independent. We observe first that for any $\psi \in \mathcal{H}$, if $\langle T_\omega\varphi_i, \psi\rangle = 0, i = 1, 2$, then $\langle\varphi_i, \gamma\rangle = 0, i = 1, 2$, for any $\gamma \in p^{-1}(P[\psi])$, and thus $\langle T_\omega(\varphi_1 + \varphi_2), \psi\rangle = 0$. Hence

$$T_\omega(\varphi_1 + \varphi_2) = z_1 T_\omega\varphi_1 + z_2 T_\omega\varphi_2$$

for some $z_1, z_2 \in \mathbb{C}$. Since φ_1, φ_2 are linearly independent there are two uniquely defined vectors θ_1, θ_2 in $[\varphi_1, \varphi_2]$, the subspace generated by the vectors φ_1, φ_2, such that $\langle\theta_i, \varphi_j\rangle = \delta_{ij}, i, j = 1, 2$. In fact, they are

$$\theta_i = \big(\langle\varphi_j,\varphi_j\rangle\varphi_i - \langle\varphi_j,\varphi_i\rangle\varphi_j\big)/\big(\langle\varphi_j,\varphi_j\rangle\langle\varphi_i,\varphi_i\rangle - \langle\varphi_i,\varphi_j\rangle\langle\varphi_j,\varphi_i\rangle\big),$$

$i = 1, 2, i \neq j$. Writing $\varphi = \varphi_1 + \varphi_2$,

$$1 = \langle\varphi, \theta_i\rangle = |\langle\varphi, \theta_i\rangle|^2 = |\langle T_\omega\varphi, T_\omega\theta_i\rangle|^2 = |z_i|^2$$

so that that $|z_i| = 1$. Since $\varphi_1, \varphi_2, \varphi \in \mathcal{O}_\omega$ and $T_\omega\varphi_1, T_\omega\varphi_2, T_\omega\varphi \in \mathcal{O}_{\omega'}$ one has $\langle\omega, \varphi\rangle = |\langle\omega, \varphi\rangle| = |\langle\omega', T_\omega\varphi\rangle| = \langle\omega', T_\omega\varphi\rangle$, which gives

$$\langle\omega, \varphi_1\rangle + \langle\omega, \varphi_2\rangle = z_1\langle\omega, \varphi_1\rangle + z_2\langle\omega, \varphi_2\rangle. \tag{2.17}$$

But then

$$\begin{aligned}\langle\omega, \varphi_1\rangle + \langle\omega, \varphi_2\rangle &= \big|\langle\omega, \varphi_1\rangle + \langle\omega, \varphi_2\rangle\big| \\ &= \big|z_1\langle\omega, \varphi_1\rangle + z_2\langle\omega, \varphi_2\rangle\big| \\ &\leq \big|z_1\langle\omega, \varphi_1\rangle\big| + \big|z_2\langle\omega, \varphi_2\rangle\big| \\ &= \langle\omega, \varphi_1\rangle + \langle\omega, \varphi_2\rangle,\end{aligned}$$

which shows that $z_1\langle\omega, \varphi_1\rangle = \lambda z_2\langle\omega, \varphi_2\rangle$ for some $\lambda \in \mathbb{R}$. Therefore, $0 < z_1\langle\omega, \varphi_1\rangle + z_2\langle\omega, \varphi_2\rangle = (1+\lambda)z_2\langle\omega, \varphi_2\rangle$, which shows that the imaginary part of z_2 equals 0 and one thus has $z_2 = \pm 1$. Similarly, one gets $z_1 = \pm 1$. From (2.17), where $\langle\omega, \varphi_1\rangle, \langle\omega, \varphi_2\rangle > 0$, one finally gets $z_1 = z_2 = 1$. This completes the proof of the additivity of T_ω.

Part 2. Let $\psi \in \mathcal{H}, \psi \neq 0$, and assume that T is any function $\mathcal{O}_\psi \to \mathcal{H}$, having the properties (2.12)–(2.15). Then for any $\varphi \in \mathcal{O}_\omega \cap \mathcal{O}_\psi$,

$$T(\varphi) = zT_\omega(\varphi), \tag{2.18}$$

for some $z \in \mathbb{T}$. Indeed, by the property (d), it holds that for any $\varphi \in \mathcal{O}_\omega \cap \mathcal{O}_\psi$, $T\varphi = f(\varphi)T_\omega\varphi$, with $f(\varphi) \in \mathbb{T}$, and it remains to be shown that $f(\varphi)$ is constant on $\mathcal{O}_\omega \cap \mathcal{O}_\psi$. For any $\lambda > 0$ and $\varphi \in \mathcal{O}_\omega \cap \mathcal{O}_\psi$, $T(\lambda\varphi) = f(\lambda\varphi)T_\omega(\lambda\varphi) = \lambda f(\lambda\varphi)T_\omega\varphi$ and $T(\lambda\varphi) = \lambda T\varphi = \lambda f(\varphi)T_\omega\varphi$. Hence $\lambda f(\lambda\varphi)T_\omega\varphi = \lambda f(\varphi)T_\omega\varphi$. Since $T_\omega\varphi \neq 0$ for $\varphi \neq 0$, this gives

$f(\varphi) = f(\lambda\varphi)$. Consider next vectors $\varphi_1, \varphi_2 \in \mathcal{O}_\omega \cap \mathcal{O}_\psi$ such that $\varphi_1 \neq \lambda\varphi_2$ for any $\lambda > 0$ (so that φ_1, φ_2 are linearly independent over $\mathbb{C}$). Then $T(\varphi_1 + \varphi_2) = f(\varphi_1 + \varphi_2)T_\omega(\varphi_1 + \varphi_2) = f(\varphi_1)T_\omega\varphi_1 + f(\varphi_2)T_\omega\varphi_2$. Using again the above vectors θ_1, θ_2, associated with φ_1, φ_2 one easily gets, e.g., $f(\varphi_1 + \varphi_2) = f(\varphi_1)$ for any $\varphi_2 \in \mathcal{O}_\omega \cap \mathcal{O}_\psi$. Hence $f(\varphi)$ is constant on $\mathcal{O}_\omega \cap \mathcal{O}_\psi$ and thus T_ω is unique modulo a phase on the cone $\mathcal{O}_\omega$.

Part 3. Let $\omega \in \mathcal{H}, \omega \neq 0$, and let $T_\omega : \mathcal{O}_\omega \to \mathcal{O}_{\omega'}$ be defined as in Part 1. We show next that T_ω has one of the following two properties, either

$$\langle T_\omega\varphi_1, T_\omega\varphi_2\rangle = \langle\varphi_1, \varphi_2\rangle \tag{2.19}$$

for all $\varphi_1, \varphi_2 \in \mathcal{O}_\omega$, or

$$\langle T_\omega\varphi_1, T_\omega\varphi_2\rangle = \langle\varphi_2, \varphi_1\rangle \tag{2.20}$$

for all $\varphi_1, \varphi_2 \in \mathcal{O}_\omega$. First of all, let $\varphi_1, \varphi_2 \in \mathcal{O}_\omega$. Then $\langle T_\omega(\varphi_1 + \varphi_2), T_\omega(\varphi_1 + \varphi_2)\rangle = \langle\varphi_1 + \varphi_2, \varphi_1 + \varphi_2\rangle$. Using the additivity of T_ω and the inner product this shows, in view of (2.16), that either $\langle T_\omega\varphi_1, T_\omega\varphi_2\rangle = \langle\varphi_1, \varphi_2\rangle$ or $\langle T_\omega\varphi_1, T_\omega\varphi_2\rangle = \langle\varphi_2, \varphi_1\rangle$. We show next that for a fixed $\varphi \in \mathcal{O}_\omega$, either $\langle T_\omega\varphi, T_\omega\psi\rangle = \langle\varphi, \psi\rangle$ or $\langle T_\omega\varphi, T_\omega\psi\rangle = \langle\psi, \varphi\rangle$ for all $\psi \in \mathcal{O}_\omega$. To prove this assume on the contrary that there are vectors $\varphi_1, \varphi_2 \in \mathcal{O}_\omega$ such that $\langle T_\omega\varphi, T_\omega\varphi_1\rangle = \langle\varphi, \varphi_1\rangle(\neq \langle\varphi_1, \varphi\rangle)$ and $\langle T_\omega\varphi, T_\omega\varphi_2\rangle = \langle\varphi_2, \varphi\rangle(\neq \langle\varphi, \varphi_2\rangle)$. By a direct computation of $\langle T_\omega\varphi, T_\omega(\varphi_1 + \varphi_2)\rangle$ one observes that this leads to a contradiction. By a similar counter argument one finally shows that either $\langle T_\omega\varphi, T_\omega\psi\rangle = \langle\varphi, \psi\rangle$ for all $\varphi, \psi \in \mathcal{O}_\omega$ or $\langle T_\omega\varphi, T_\omega\psi\rangle = \langle\psi, \varphi\rangle$ for all $\psi \in \mathcal{O}_\omega$.

Part 4. We construct next a unitary or antiunitary operator U of $\mathcal{H}$ for which $p(P) = UPU^*$ for all $P \in \mathbf{P}$.

Let $\omega \in \mathcal{H}$ and $T_\omega : \mathcal{O}_\omega \to \mathcal{O}_{\omega'}$ be given as in Part 1. Let $M = [\omega]^\perp$ and $M' = [\omega']^\perp$ and define a function $S : M \to M'$ by

$$\begin{aligned} S\varphi &:= T_{\omega+\varphi}\varphi, \quad && \varphi \neq 0 \\ S\varphi &:= 0, && \varphi = 0 \end{aligned}$$

where $T_{\omega+\varphi}$ is the operator on the cone $\mathcal{O}_{\omega+\varphi}$ with the choice of the phase given by $T_{\omega+\varphi}\omega = \omega'$. S is well defined since for any $\varphi \in M$, $\varphi \neq 0$, we have $\varphi \in \mathcal{O}_{\omega+\varphi}$. Moreover, for any two $\varphi, \psi \in M$, $T_{\omega+\varphi} = T_{\omega+\psi}$ on the cone $\mathcal{O}_{\omega+\varphi} \cap \mathcal{O}_{\omega+\psi}$, which contains at least the vector ω for which $T_{\omega+\varphi}\omega = T_{\omega+\psi}\omega$. According to Part 3 any $T_{\omega+\varphi}$, $\varphi \in M$, has either the property (2.19) or the property (2.20). Due to the fact that for all $\varphi, \psi \in M$, $T_{\omega+\varphi} = T_{\omega+\psi}$ on the intersection of their defining cones, all the operators $T_{\omega+\varphi}, \varphi \in M$, are of the type (2.19) or they all are of the type (2.20). We proceed to show that S is in the first case a unitary operator and in the second case an antiunitary operator. In fact the proofs of the two different cases are similar and we treat only the case that all $T_{\omega+\varphi}, \varphi \in M$, are of the type (2.19).

We show first that for any $\varphi \in M, \lambda \in \mathbb{C}$, $S(\lambda\varphi) = \lambda S\varphi$. In fact, if $\lambda\varphi = 0$, the result is obvious, otherwise we have

$$\begin{aligned}\langle T_\omega(\omega+\lambda\varphi), T_\omega(\omega+\varphi)\rangle &= \langle \omega+\lambda\varphi, \omega+\varphi\rangle \\ &= \|\omega\|^2 + \bar{\lambda}\langle\varphi,\varphi\rangle \\ \langle T_\omega(\omega+\lambda\varphi), T_\omega(\omega+\varphi)\rangle &= \langle T_{\omega+\lambda\varphi}(\omega+\lambda\varphi), T_{\omega+\varphi}(\omega+\varphi)\rangle \\ &= \langle T_{\omega+\lambda\varphi}\omega + T_{\omega+\lambda\varphi}(\lambda\varphi), T_{\omega+\varphi}\omega + T_{\omega+\varphi}\varphi\rangle \\ &= \langle \omega' + S(\lambda\varphi), \omega' + S\varphi\rangle \\ &= \|\omega'\|^2 + \langle S(\lambda\varphi), S\varphi\rangle.\end{aligned}$$

Since $\|\omega\| = \|\omega'\|$ this gives $\langle S(\lambda\varphi), S\varphi\rangle = \bar{\lambda}\langle\varphi,\varphi\rangle$. But $S(\lambda\varphi) = T_{\omega+\lambda\varphi}(\lambda\varphi) \in p(P[\lambda]\varphi)$ and $S\varphi \in p(P[\varphi])$, which shows that $S(\lambda\varphi) = zS\varphi$ for some $z \in \mathbb{C}$. Therefore, $\bar{\lambda}\langle\varphi,\varphi\rangle = \langle S(\lambda\varphi), S\varphi\rangle = \bar{z}\langle S\varphi, S\varphi\rangle = \bar{z}\langle\varphi,\varphi\rangle$, which gives $\bar{z} = \bar{\lambda}$, and thus $S(\lambda\varphi) = \lambda S\varphi$.

To show the additivity of S on M, let $\varphi_1, \varphi_2 \in M$. If $\varphi_1 = \lambda\varphi_2, \lambda \in \mathbb{C}$, then the homogeneity of S gives the additivity. Therefore, assume that φ_1, φ_2 are linearly independent. Let θ_1, θ_2 be the unique vectors in $[\varphi_1, \varphi_2]$ such that $\langle\theta_i, \varphi_j\rangle = \delta_{ij}$. Then

$$\begin{aligned}S(\varphi_1+\varphi_2) &= T_{\omega+\varphi_1+\varphi_2}(\varphi_1+\varphi_2) \\ &= T_{\omega+\theta_1+\theta_2}(\varphi_1+\varphi_2) \\ &= T_{\omega+\theta_1+\theta_2}\varphi_1 + T_{\omega+\theta_1+\theta_2}\varphi_2 \\ &= T_{\omega+\varphi_1}\varphi_1 + T_{\omega+\varphi_2}\varphi_2 = S\varphi_1 + S\varphi_2.\end{aligned}$$

Hence $S : M \to M'$ is a linear map. It is also isometric since for any $\varphi \in M$, $\varphi \neq 0$, $\langle S\varphi, S\varphi\rangle = \langle T_{\omega+\varphi}\varphi, T_{\omega+\varphi}\varphi\rangle = \langle T_\varphi\varphi, T_\varphi\varphi\rangle = \langle\varphi,\varphi\rangle$. Moreover, for any unit vector $\varphi \in M$ one has $P[S]\varphi = p(P[\varphi])$. To show the surjectivity of S, let $\psi \in M', \psi \neq 0$. Since p is surjective there is a unit vector $\varphi \in M$ such that $p(P[\varphi]) = P[\psi]$. Hence $S\varphi = \lambda\psi$ for some $\lambda \in \mathbb{C}$. Since $\|\varphi\| = 1$, also $\|S\varphi\| = 1$ so that $\lambda \neq 0$ and thus $S(\frac{\varphi}{\lambda}) = \psi$. This concludes the proof of the unitarity of S.

We now have $\mathcal{H} = [\omega] \oplus M = [\omega'] \oplus M'$ and we define $U : \mathcal{H} \to \mathcal{H}$ such that $U(\lambda\omega + \varphi) = \lambda\omega' + S\varphi$ for all $\lambda \in \mathbb{C}, \varphi \in M$. If S is antiunitary we define U instead by $U(\lambda\omega+\varphi) = \bar{\lambda}\omega' + S\varphi$. Clearly, the operator U is unitary (antiunitary) and it is related to the function p according to $p(P) = UPU^*$ for any $P \in \mathbf{P}$.

Part 5. Let $V : \mathcal{H} \to \mathcal{H}$ be related to p according to $p(P) = VPV^*$, $P \in \mathbf{P}$. By change of phase we may assume that $V\omega = \omega'$. Let $\varphi \in M$. The operator V has, in particular, the properties (2.12)–(2.15) on $\mathcal{O}_{\omega+\varphi}$ so that V, when restricted on $\mathcal{O}_{\omega+\varphi}$, equals $zT_{\omega+\varphi}$ for some $z \in \mathbb{T}$. But since $V\omega = \omega' = zT_{\omega+\varphi}\omega = z\omega'$, one has that for any $\varphi \in M$, $V|_{\mathcal{O}_{\omega+\varphi}} = T_{\omega+\varphi}$, that is, $V\varphi = S\varphi$ on M. Therefore, V equals with U on M, showing that $V = U$ whenever $M \neq \{0\}$. In other words, U is unique modulo a phase factor and the unitary or the antiunitary nature of U is completely determined by $p \in \mathrm{Aut}\,(\mathbf{P})$ (apart from the trivial case of $\mathcal{H}$ being one-dimensional). Moreover, the operator U does not depend on the choice of the vector ω. This ends the proof of the theorem. □

The following result is an immediate corollary to the Wigner theorem.

Corollary 1. *For any* $p \in \mathrm{Aut}\,(\mathbf{P})$ *there is a unique* $\sigma_p \in \Sigma$ *such that* $p = p_{\sigma_p}$. *The function* $\mathrm{Aut}\,(\mathbf{P}) \ni p \mapsto \sigma_p \in \Sigma$ *is an injective group homomorphism.*

As another application of the Wigner theorem one gets the following from Lemma 5:

Corollary 2. *Assume that* $\dim(\mathcal{H}) \geq 2$. *For any* $e \in \mathrm{Aut}_o(\mathbf{E})$ *there is a unique* $\sigma_e \in \Sigma$ *such that* $e = e_{\sigma_e}$.

2.3 The Group Isomorphisms

2.3.1 Isomorphisms

In this section we collect the results obtained in the previous sections on the relations between the various automorphism groups of quantum mechanics. This will allow us to prove, apart from the particular cases of one and two dimensional Hilbert spaces, that all these groups are isomorphic.

Consider the following diagrams:

$$\begin{array}{c} \dim(\mathcal{H}) \geq 1 \\ \\ \mathrm{Aut}\,(\mathbf{D}) \\ \updownarrow \\ \mathrm{Aut}_0(\mathbf{P}) \end{array} \qquad \begin{array}{ccc} & \dim(\mathcal{H}) \geq 1 & \\ & & \\ \mathrm{Aut}\,(\mathbf{S}) & \xleftarrow{4} & \mathrm{Aut}_s(\mathbf{E}) \\ \downarrow{\scriptstyle 1} & & \uparrow{\scriptstyle 3} \\ \mathrm{Aut}\,(\mathbf{P}) & \xrightarrow{2} & \Sigma \end{array} \qquad \begin{array}{c} \dim(\mathcal{H}) \geq 2 \\ \\ \mathrm{Aut}_o(\mathbf{E}) \\ \updownarrow \\ \Sigma \end{array}$$

The double arrow in the first diagram is the isomorphism of Proposition 7, while the arrows 1 to 4 in the second diagram are injective homomorphisms given by Proposition 1, Corollary 1, Proposition 9 and Proposition 4, respectively. Proposition 9 and Corollary 2 give the double arrow in the third diagram. We show that the map obtained by composing the arrows in the second diagram is the identity. From this it follows that the maps involved are isomorphisms.

Let G denote any of the four groups appearing in the second diagram. Starting from G and composing the injective group homomorphisms, one obtains an injective group homomorphism ϕ_G of G into G.

Corollary 3. *The map* ϕ_G *is the identity on* G.

Proof. It is sufficient to prove the statement for a particular choice of G. Choosing, for instance, $G = \mathrm{Aut}\,(\mathbf{S})$, we then have

$$\phi_{\mathrm{Aut}\,(\mathbf{S})} : s \mapsto p_s \mapsto \sigma_{p_s} \mapsto e_{\sigma_{p_s}} \mapsto s_{e_{\sigma_{p_s}}},$$

where we use the notations $p_s := s|_{\mathbf{P}}$ and $e_\sigma := g_\sigma|_{\mathbf{E}}$. It can immediately be checked that $\phi_{\mathrm{Aut}\,(\mathbf{S})}(s) = s$. □

Now suppose that the dimension of the Hilbert space is at least three and consider the following diagram:

$$\begin{array}{ccccc} \mathrm{Aut}_s(\mathbf{P}) & \xrightarrow{6} & \mathrm{Aut}_0(\mathbf{P}) & \longrightarrow & \mathrm{Aut}\,(\mathbf{D}) \\ \uparrow 5 & & & & \downarrow 7 \\ \Sigma & \longleftarrow & \mathrm{Aut}\,(\mathbf{P}) & \longleftarrow & \mathrm{Aut}\,(\mathbf{S}) \end{array}$$

Here the new arrows 5, 6 and 7 are given, respectively, by Proposition 9, the natural immersion $\mathrm{Aut}_s(\mathbf{P}) \hookrightarrow \mathrm{Aut}_0(\mathbf{P})$, which is obviously an injective group homomorphism, and by Proposition 8.

As in Corollary 3, one can prove that the loop contained in the previous diagram is in fact an identity. Hence we conclude that, if $\dim \mathcal{H} \geq 3$, all the groups considered are isomorphic. We summarize this fact in the following corollary.

Corollary 4. *If* $\dim(\mathcal{H}) \geq 3$, *the groups* Σ, $\mathrm{Aut}\,(\mathbf{S})$, $\mathrm{Aut}_s(\mathbf{P})$, $\mathrm{Aut}\,(\mathbf{P})$, $\mathrm{Aut}_0(\mathbf{P})$, $\mathrm{Aut}_s(\mathbf{E})$, $\mathrm{Aut}_c(\mathbf{E})$, $\mathrm{Aut}_o(\mathbf{E})$ *and* $\mathrm{Aut}\,(\mathbf{D})$ *are isomorphic.*

Remark 3. As pointed out in Remark 2, if $\dim(\mathcal{H}) \geq 3$, then the structure of bijective maps $e : \mathbf{E} \to \mathbf{E}$, which preserve both the order and the coexistence of effects, can also be determined and they are of the form $e = e_U$, $U \in \mathrm{Aut}\,(\mathcal{H})$, [30]. The list of Corollary 4 could thus be extended further with another group of automorphisms.

2.3.2 Homeomorphisms

We proceed to show that the isomorphisms of the previous section are homeomorphisms when the groups are equipped with their natural topologies induced by the duality $(T, E) \mapsto \mathrm{tr}\left[TE\right]$. Let $\mathbf{X}$ denote one of the sets $\mathbf{S}$, $\mathbf{P}$, $\mathbf{E}$, or $\mathbf{D}$, and let $G(\mathbf{X})$ stand for any of the groups of automorphisms of $\mathbf{X}$ considered so far, endowed with its natural topology. According to the restrictions on the dimension of $\mathcal{H}$ imposed by Corollary 4 we can suppose that the dimension of $\mathcal{H}$ is greater than 2. Hence, in each case the group $G(\mathbf{X})$ is isomorphic to Σ and for any $x \in G(\mathbf{X})$ there is a unique $\sigma_x \in \Sigma$ such that $x = x_{\sigma_x}$, with $x_{\sigma_x} = g_{\sigma_x}|_{\mathbf{X}}$. Thus for any $U \in \sigma_x$, $x = x_U$, that is,

$$x(A) = x_U(A) = UAU^*, \quad A \in \mathbf{X}.$$

Proposition 10. *The map* $j_{\mathbf{X}} : \Sigma \to G(\mathbf{X}), \sigma \mapsto x_\sigma$, *is a group homeomorphism, and* $G(\mathbf{X})$ *is a second countable, metrisable, topological group.*

Proof. The group Σ is a second countable, metrisable, topological group (see Appendix A.2) and the map $j_{\mathbf{X}}$ is bijective so that it remains to be shown that it is a homeomorphism. We demonstrate first that the function $J_{\mathbf{X}} : \mathbf{U} \cup \overline{\mathbf{U}} \to G(\mathbf{X}), U \mapsto J_{\mathbf{X}}(U) := x_U$ is continuous. Since $\mathbf{U} \cup \overline{\mathbf{U}}$ is second countable, it suffices to show that if $(U_n)_{n\geq 1}$ is a (strongly) convergent sequence in $\mathbf{U} \cup \overline{\mathbf{U}}$, then $(J_{\mathbf{X}}(U_n))_{n\geq 1}$ is convergent in $G(\mathbf{X})$. As $U \mapsto U^{-1}$ is continuous in $\mathbf{U} \cup \overline{\mathbf{U}}$, we have, for instance for $\mathbf{X} = \mathbf{S}$,

$$\begin{aligned}\lim_{n\to\infty} f_{T,E}(J_{\mathbf{S}}(U_n)) &= \lim_{n\to\infty} \mathrm{tr}\big[J_{\mathbf{S}}(U_n)(T)E\big] \\ &= \lim_{n\to\infty} \mathrm{tr}\big[U_n T U_n^{-1} E\big] \\ &= \mathrm{tr}\big[UTU^{-1}E\big] = f_{T,E}(J_{\mathbf{S}}(U)),\end{aligned}$$

for all $E \in \mathbf{E}, T \in \mathbf{S}$, which shows the continuity of $J_{\mathbf{S}}$. The other cases are shown as well. By definition of quotient topology, this also proves that $j_{\mathbf{X}}$ is continuous. It remains to be shown that the inverse mapping $j_{\mathbf{X}}^{-1}$ is continuous. Consider the group $G(\mathbf{X})$ and let $(\varphi_i)_{i\geq 1}$ be a dense sequence of unit vectors in $\mathcal{H}$. Since $\mathbf{P}$ is contained in $\mathbf{X}$, the sequence of functions $\big(f_{P[\varphi_i],P[\varphi_j]}\big)_{i,j\geq 1}$ gives $G(\mathbf{X})$ a metrisable topology, which a priori is weaker than the one defined above for $G(\mathbf{X})$. We shall show that $j_{\mathbf{X}}^{-1}$ is continuous in this weaker topology. It suffices again to consider only sequences. Let (x_n) be a convergent sequence in $G(\mathbf{X})$, with $x_n \to x$. We will show that $j_{\mathbf{X}}^{-1}(x_n) \to j_{\mathbf{X}}^{-1}(x)$ in Σ. To proceed assume on the contrary that $j_{\mathbf{X}}^{-1}$ is not continuous, so that there is an open set $O \subset \Sigma$ such that $j_{\mathbf{X}}^{-1}(x) \in O$ but $j_{\mathbf{X}}^{-1}(x_{n_k}) \notin O$ for a subsequence (x_{n_k}) of (x_n). Let $U_k, U \in \mathbf{U} \cup \overline{\mathbf{U}}$ such that $j_{\mathbf{X}}([U_k]) = x_{n_k}$ and $j_{\mathbf{X}}([U]) = x$. The sequence (U_k) is bounded, so that it has a weakly convergent subsequence (U_{k_h}) in $\mathbf{U} \cup \overline{\mathbf{U}}$, with $U_{k_h} \to V$. But then $\mathrm{tr}\big[P[\varphi_i]x_{n_{k_h}}(P[\varphi_j])\big] = |\langle\varphi_i, U_{n_{k_h}}\varphi_j\rangle|^2 \to |\langle\varphi_i, V\varphi_j\rangle|^2$ and $\mathrm{tr}\big[P[\varphi_i]x_{n_{k_h}}(P[\varphi_j])\big] \to \mathrm{tr}\big[P[\varphi_i]x(P[\varphi_j])\big] = |\langle\varphi_i, U\varphi_j\rangle|^2$, which shows that $[V] = [U]$. Since $U_{n_{k_h}} \to V$ also strongly we thus have $[U_{n_{k_h}}] \to [V] = [U]$, which is a contradiction. This shows that $j_{\mathbf{X}}^{-1} : G(\mathbf{X}) \to \Sigma$ is continuous. This ends the proof. □

2.3.3 The Automorphism Group of Quantum Mechanics

On the basis of Corollary 4 and Proposition 10 all the groups considered so far are isomorphic and homeomorphic with each other in a natural way, with the dimension requirement $\dim(\mathcal{H}) \geq 2$ for the group $\mathrm{Aut}_o(\mathbf{E})$, and $\dim(\mathcal{H}) \geq 3$ for the groups $\mathrm{Aut}\,(\mathbf{D})$, $\mathrm{Aut}_0(\mathbf{P})$, and $\mathrm{Aut}_s(\mathbf{P})$. Any of these groups may thus be called the automorphism group of quantum mechanics. The implementation of the structure preserving transformations in terms of unitary or antiunitary operators is, however, most directly obtained from the group Σ in terms of a section $s : \Sigma \to \mathbf{U} \cup \overline{\mathbf{U}}$ for the canonical projection $\pi : \mathbf{U} \cup \overline{\mathbf{U}} \to \Sigma$. Therefore, from now on we shall refer to the group Σ as *the automorphism group of quantum mechanics.*

3 The Symmetry Actions and Their Representations

This chapter is devoted to the study of the homomorphisms

$$G \ni g \mapsto \sigma_g \in \Sigma,$$

where G is a connected Lie group and Σ the symmetry group of quantum mechanics. We call such a homomorphism *symmetry action*, leaving the word "representation" for the more specific use of representing a group in terms of unitary or antiunitary operators on the underlying Hilbert space. It is the notion of symmetry action which formalizes the idea that a group G is a symmetry group of a quantum system described by a Hilbert space $\mathcal{H}$. The assumption that G is a Lie group is satisfied by most groups of physical relevance, like the Euclidean group, the Galilei group, or the Poincaré group. The assumption on the connectedness of G implies that we consider only continuous symmetries. These mathematical assumptions are crucial in determining the symmetry actions of a group.

We will show that there is a natural connection between symmetry actions of a group G and representations of another group $\overline{G}$, the universal central extension of G, and it is this connection which is of primary importance in the physical applications.

The material of this chapter is organized in the following way. Section 3.1 introduces the basic definitions concerning the symmetry actions of a Lie group. These definitions do not depend on the Lie structure of the group. Section 3.2 collects some technical results on the multipliers of a connected simply connected Lie group. Section 3.3 reviews the construction of the universal central extension for a connected Lie group and presents the fundamental connection between the symmetry actions of such a group and the unitary representations of its universal central extension. In Sect. 3.4 the Mackey Machine of Appendix A.3 will be applied to construct these representations for the case where the universal central extension of the symmetry group is a regular semidirect product. Temporal evolution of a closed system will offer the first elementary example of the application of the general theory (Sect. 3.5).

G. Cassinelli, E. De Vito, P.J. Lahti, and A. Levrero, *The Theory of Symmetry Actions in Quantum Mechanics*, Lect. Notes Phys. **654**, pp. 27–47
http://www.springerlink.com/

3.1 Symmetry Actions of a Lie Group

Let G be a topological group and $\mathcal{H}$ a complex separable Hilbert space. The following definition translates in mathematically precise language the physical requirement that G is a group of symmetries for a quantum system represented by the Hilbert space $\mathcal{H}$.

Definition 8. A function $G \ni g \mapsto \sigma_g \in \Sigma$ is a *symmetry action* of G on $\mathcal{H}$ if it is a continuous group homomorphism, that is, a continuous function having the properties $\sigma_e = [I]$ and $\sigma_{g_1 g_2} = \sigma_{g_1}\sigma_{g_2}$ for all $g_1, g_2 \in G$.

According to Proposition 12 and the von Neumann theorem (Lemma 14, A.1), the continuity of a symmetry action $\sigma : G \to \Sigma$ is already implied by its measurability. This is a useful result since the measurability of σ is often simpler to verify than its continuity.

Let G be a group and consider the problem of describing all the quantum systems that have G as a group of symmetries. To solve this problem in an appropriate way one must take into account the fact that physics does not fix in a unique way the Hilbert space associated with a quantum system. Therefore, let $\mathcal{H}$ and $\mathcal{H}'$ be two Hilbert spaces and define $\Sigma(\mathcal{H}, \mathcal{H}')$ as the set of equivalence classes of unitary or antiunitary operators $B : \mathcal{H} \to \mathcal{H}'$ with respect to the following relation: B_1 is equivalent to B_2 if there is a $z \in \mathbb{T}$ such that $B_1 = zB_2$. Let $[B] := \{zB \mid z \in \mathbb{T}\}$ denote the equivalence class of B and extend the operator product to these classes such that if $[B_1] \in \Sigma(\mathcal{H}, \mathcal{H}')$ and $[B_2] \in \Sigma(\mathcal{H}', \mathcal{H}'')$, then $[B_1][B_2] := [B_1 B_2] \in \Sigma(\mathcal{H}, \mathcal{H}'')$. Clearly, each element of $\beta \in \Sigma(\mathcal{H}, \mathcal{H}')$ allows one to describe the quantum system associated with $\mathcal{H}$ in terms of the mathematical objects defined on $\mathcal{H}'$ in such a way that the probabilistic structure of the theory is completely preserved. In particular, any such $\beta \in \Sigma(\mathcal{H}, \mathcal{H}')$ establishes a one-to-one correspondence between the symmetry groups $\Sigma(\mathcal{H})$ and $\Sigma(\mathcal{H}')$ through the mapping

$$\Sigma(\mathcal{H}) \ni \sigma \mapsto \beta\sigma\beta^{-1} \in \Sigma(\mathcal{H}').$$

In view of the above considerations the following definition is natural.

Definition 9. Two symmetry actions $\sigma : G \to \Sigma(\mathcal{H})$ and $\sigma' : G \to \Sigma(\mathcal{H}')$ of a group G on the Hilbert spaces $\mathcal{H}$ and $\mathcal{H}'$, respectively, are *equivalent* if there is a $\beta \in \Sigma(\mathcal{H}, \mathcal{H}')$ such that $\beta\sigma_g = \sigma'_g\beta$ for all $g \in G$.

Different physical systems which behave in the same way under the action of a group G are thus characterized by the property that the corresponding symmetry actions are equivalent.

We say that a physical system is *elementary* with respect to the symmetry action $\sigma : G \to \Sigma$ if for any vector state $P \in \mathbf{P}$ the set $\{\sigma_g(P) \mid g \in G\}$ of vector states is complete in the sense of superpositions, that is, any other vector state $P_1 \in \mathbf{P}$ can be expressed as a superposition of some of the vector states $\sigma_g(P), g \in G$ (see the discussion at the beginning of Chap. 1.1). This is equivalent with the fact that for any $P \in \mathbf{P}$ the least upper bound of the projections $\sigma_g(P)$, $g \in G$, is the identity operator, that is, $\vee_{g\in G}\sigma_g(P) = I$.

Lemma 6. *Let $\sigma : G \to \Sigma$ be a symmetry action. The following two conditions are equivalent:*

a) for any $P \in \mathbf{P}$, $\vee_{g\in G}\sigma_g(P) = I$;
b) for any $P_1, P_2 \in \mathbf{P}$ there is a $g \in G$ such that $\mathrm{tr}\big[P_1\sigma_g(P_2)\big] \neq 0$.

Proof. Assume a) and let $P_1, P_2 \in \mathbf{P}$, so that, for instance, $\vee_{g\in G}\sigma_g(P_2) = I$. If $\mathrm{tr}\big[P_1\sigma_g(P_2)\big] = 0$ for all $g \in G$, then $\sigma_g(P_2) \leq P_1^\perp$ for all $g \in G$. Therefore, $I = \vee_{g\in G}\sigma_g(P_2) \leq P_1^\perp$, so that $P_1 = 0$, which is a contradiction. Assume now b), and choose a $P \in \mathbf{P}$. Assume that the projection $R = \vee_{g\in G}\sigma_g(P_2)$ is not I, so that its complement $R^\perp$ is nonzero. Let Q be any one-dimensional projection contained in $R^\perp$. Then $R \leq Q^\perp$, and thus $\sigma_g(P) \leq Q^\perp$ for all $g \in G$. But then $\mathrm{tr}\big[Q\sigma_g(P)\big] = 0$ for any $g \in G$, which conflicts with b).

This lemma leads to the following physically motivated notion of the irreducibility of a symmetry action. Its mathematical correctness becomes clear in Theorem 3.

Definition 10. A symmetry action $\sigma : G \to \Sigma$ is *irreducible* if for any $P_1, P_2 \in \mathbf{P}$ there is a $g \in G$ such that $\mathrm{tr}\big[P_1\sigma_g(P_2)\big] \neq 0$.

The classification of the possible elementary quantum systems having G as the symmetry group is traced back to the mathematical problem of finding all the irreducible symmetry actions of G, up to an equivalence. To this aim an essential step is to study the connection between the symmetry actions and the unitary / antiunitary representations.

From now on we assume that G is a connected Lie group. Since $\Sigma_0 := \mathbf{U}/\mathbf{T}$ is the connected component of the identity of Σ and the map $g \mapsto \sigma_g$ is continuous, then, for all $g \in G$, $\sigma_g \in \Sigma_0$, that is, all the symmetries σ_g are induced by unitary operators.

Let $U : G \to \mathbf{U}$ be a unitary representation of G and let $\pi : \mathbf{U} \to \Sigma_0$ be the canonical projection. Then the map

$$G \ni g \mapsto \sigma_g := \pi(U_g) \in \Sigma_0$$

is a symmetry action of G on $\mathcal{H}$. Furthermore, if U and U' are unitarily equivalent unitary representations of G in $\mathcal{H}$ and $\mathcal{H}'$, respectively, the symmetry actions $G \ni g \mapsto \pi(U_g) \in \Sigma_0(\mathcal{H})$ and $G \ni g \mapsto \pi(U'_g) \in \Sigma_0(\mathcal{H}')$ are equivalent as well. Moreover, if $g \mapsto U_g$ is an irreducible representation, then also $g \mapsto \pi(U_g)$ is irreducible.

The unitary representations of G are not enough to describe all the symmetry actions of G, and there are unitarily inequivalent representations of G such that the corresponding symmetry actions are equivalent. Indeed, let $\sigma : G \to \Sigma_0$ be a symmetry action and $s : \Sigma_0 \to \mathbf{U}$ a measurable section for the canonical projection $\pi : \mathbf{U} \to \Sigma_0$, that is, a measurable function such that $\pi(s(\sigma)) = \sigma$ for all $\sigma \in \Sigma_0$. Then the function

$$G \ni g \mapsto U_g := s(\sigma_g) \in \mathbf{U}$$

is measurable, $U_e = I$, but, instead of $U_{g_1 g_2} = U_{g_1} U_{g_2}$, one only gets the weaker condition

$$U_{g_1 g_2} = z(g_1, g_2) U_{g_1} U_{g_2}, \tag{3.1}$$

with $z(g_1, g_2) \in \mathbb{T}$. The fact that σ is a group homomorphism implies that

$$z(g_1 g_2, g_3) z(g_1, g_2) = z(g_1, g_2 g_3) z(g_2, g_3) \tag{3.2}$$

$$z(g, e) = z(e, g) = 1. \tag{3.3}$$

The map $(g_1, g_2) \mapsto z(g_1, g_2)$ satisfying (3.2) and (3.3) is called a $\mathbb{T}$-multiplier and the map $g \mapsto U_g$ satisfying (3.1) is known as a *projective representation* of G, with the multiplier z. Moreover, if σ' is another symmetry action equivalent to σ and acting in $\mathcal{H}'$, then, by definition, there is a unitary or antiunitary operator B from $\mathcal{H}$ onto $\mathcal{H}'$ and a measurable function $G \to \mathbb{T}$ such that

$$U'_g = b(g) B U_g B^{-1} \qquad g \in G,$$

where $U'_g := s(\sigma'_g)$. Conversely, given a projective representation $g \mapsto U_g$ of G, $g \mapsto \pi(U_g)$ is a symmetry action of G. Moreover, if $g \mapsto U'_g$ is another projective representation of G such that there is a unitary or antiunitary operator B from $\mathcal{H}$ onto $\mathcal{H}'$ and a measurable function $b : G \to \mathbb{T}$ such that $U'_g = b(g) B U_g B^{-1}$ for all $g \in G$, then the symmetry actions $g \mapsto \pi(U_g)$ and $g \mapsto \pi(U'_g)$ are equivalent.

The problem of determining the symmetry actions of G is reduced to the study of the projective representations of G and of finding a suitable notion of equivalence that generalizes the unitary equivalence of representations. In order to apply the powerful theory of (ordinary) representations one needs to take a second step. This consists of defining a group $\overline{G}$ such that its irreducible unitary representations are in one to one correspondence with the irreducible projective representations of G. Such a group $\overline{G}$ will be constructed in Sect. 3.3 and it will be called the universal central extension of G. Its construction depends heavily on the structure of the set of the $\mathbb{T}$-multipliers of G.

The determination of the $\mathbb{T}$-multipliers of a Lie group is, in general, a highly difficult nonlinear problem. However, the classification of the $\mathbb{T}$-multipliers of a connected, simply connected Lie group can be reduced to a finite-dimensional linear problem on the Lie algebra of the group.

In many physical applications the group G is not simply connected. To bypass this difficulty, one can consider the universal covering group G^* of G. By definition, it is simply connected. However, in general, the set of the

$\mathbb{T}$-multipliers of G may be quite different from the set of the $\mathbb{T}$-multipliers of G^*. For example, the Poincaré group has (essentially) two multipliers (corresponding to the bosonic and the fermionic particles, respectively), whereas its universal covering group only has the trivial multiplier.

It will be shown in the following sections that, despite this fact, the study of the $\mathbb{T}$-multipliers of G^* is sufficient to classify all the symmetry actions of G, also when G is not simply connected. This remarkable mathematical fact has created some confusion in the physics literature, where, sometimes, instead of the natural symmetry group G its covering group G^* is considered as the true group of physical symmetries.

Remark 4. The common way to solve the problem of passing from projective representations to (ordinary) representations starts with classifying the $\mathbb{T}$-multipliers of G. Each multiplier z is then used to define a group G_z, the central extension of G by $\mathbb{T}$. Finally, one classifies all the irreducible representations of G_z (see, for example, [35]).

3.2 Multipliers for Lie Groups

This section gives a brief summary of the part of the theory of multipliers needed to describe the multipliers for a simply connected Lie group. The proofs of the quoted results can be read, for instance, in Chap. 7 of [35], where a systematic study of the multipliers is presented.

Let H be a connected and A a commutative Lie group, and let e and 1 be their respective unit elements.

Definition 11. An *A-multiplier* of H is a measurable map $\tau : H \times H \to A$ for which

$$\begin{aligned} \tau(e,g) &= \tau(g,e) = 1, && g \in H, \\ \tau(g_1, g_2 g_3)\tau(g_2, g_3) &= \tau(g_1, g_2)\tau(g_1 g_2, g_3), && g_1, g_2, g_3 \in H. \end{aligned}$$

Two A-multipliers τ_1 and τ_2 of H are *equivalent* if there is a measurable map $b : H \to A$ such that

$$\tau_2(g_1, g_2) = \frac{b(g_1 g_2)}{b(g_1)b(g_2)}\tau_1(g_1, g_2), \qquad g_1, g_2 \in H.$$

An A-multiplier τ is *exact* if it is equivalent to the constant multiplier 1, that is,

$$\tau(g_1, g_2) = \frac{b(g_1 g_2)}{b(g_1)b(g_2)}, \qquad g_1, g_2 \in H,$$

for some measurable map b from H to A.

The set of A-multipliers is a commutative group under pointwise multiplication and the set of exact A-multipliers is a subgroup of it. We let $H^2(H,A)$ denote the corresponding quotient group.

The introduction of the above notion of equivalence of multipliers is motivated by the following observation. Let G be a group of symmetries and assume that it is a connected Lie group. Let $\sigma : G \to \Sigma_0$ be a symmetry action of G. Given a measurable section $s : \Sigma_0 \to \mathbf{U}$ for the canonical projection $\pi : \mathbf{U} \to \Sigma_0$, define, for all $g, g_1, g_2 \in G$,

$$\begin{aligned} U_g &= s(\sigma_g) \\ z(g_1,g_2)I &= U_{g_1g_2}U_{g_2}^{-1}U_{g_1}. \end{aligned}$$

Then z is a $\mathbb{T}$-multiplier of G and U is a projective representation with z as its multiplier (compare with (3.1)). If s' is another measurable section for π and z' is the multiplier of the projective representation $g \mapsto s'(\sigma_g)$, then z and z' are equivalent $\mathbb{T}$-multipliers. Moreover, if τ is a $\mathbb{T}$-multiplier of G, there always exists a projective representation U of G having τ as its multiplier, see [35]. If τ' is another multiplier equivalent to τ, there is a projective representation having τ' as its multiplier and which induces the same symmetry action as U does.

The above observations imply that in order to classify all the symmetry actions of a group G it suffices to study the quotient group $H^2(G,\mathbb{T})$ instead of the group of the $\mathbb{T}$-multipliers of G.

From now on we assume that the generic Lie group H is simply connected.

The following lemma, see Corollary 7.32 of [35], reduces the study of the $\mathbb{T}$-multipliers of H to the study of its $\mathbb{R}$-multipliers.

Lemma 7. *Each $\mathbb{T}$-multiplier of H is equivalent to one of the form $e^{i\tau}$, where τ is an $\mathbb{R}$-multiplier of H. The multiplier τ is exact if and only if the multiplier $e^{i\tau}$ is exact.*

In general, multipliers are measurable functions. However, in the case of the real valued multipliers one may restrict the study to analytic multipliers only. This is due to the following result, see Corollary 7.30 of [35].

Lemma 8. *Any $\mathbb{R}^n$-multiplier of H is equivalent to an analytic one.*

The above two lemmas imply that each $\mathbb{T}$-multiplier is equivalent to a multiplier $e^{i\tau}$, where τ is an analytic $\mathbb{R}$-multiplier. Moreover, since τ is analytic and H is simply connected, the multipliers may be studied from an infinitesimal point of view. To do this, we need the following definition. In that $\mathrm{Lie}\,(H)$ denotes the Lie algebra of H and $(X,Y) \mapsto [X,Y]$ is its Lie product.

Definition 12. A bilinear skew symmetric map $F : \mathrm{Lie}\,(H) \times \mathrm{Lie}\,(H) \to \mathbb{R}^n$ for which

$$F(X,[Y,Z]) + F(Z,[X,Y]) + F(Y,[Z,X]) = 0, \qquad X,Y,Z \in \mathrm{Lie}\,(H),$$

is a *closed* $\mathbb{R}^n$*-form*. A closed $\mathbb{R}^n$-form F is *exact* if there is a linear map $q : \mathrm{Lie}\,(H) \to \mathbb{R}^n$ such that

$$F(X,Y) = q([X,Y]), \qquad X,Y \in \mathrm{Lie}\,(H).$$

The set of closed $\mathbb{R}^n$-forms is a finite dimensional real vector space and the set of exact $\mathbb{R}^n$-forms is a subspace of it. Let $H^2(\mathrm{Lie}\,(H), \mathbb{R}^n)$ denote the corresponding quotient space.

The above definition is motivated by the following result, which is essential in order to define the universal central extension of a connected Lie group. Observe that the set of $\mathbb{R}^n$-multipliers is a real vector space under the pointwise operations and the set of exact $\mathbb{R}^n$-multipliers is a subspace of it, so that the group $H^2(H, \mathbb{R}^n)$ is also a vector space.

Theorem 2. *The vector spaces $H^2(H,\mathbb{R}^n)$ and $H^2(\mathrm{Lie}\,(H),\mathbb{R}^n)$ are isomorphic in a canonical way.*

Proof. To exhibit the isomorphism claimed, let F be a closed $\mathbb{R}^n$-form and denote by $\mathbb{R}^n \oplus_F \mathrm{Lie}\,(H)$ the Lie algebra defined by the following Lie bracket:

$$\big[(v_1, X_1), (v_2, X_2)\big] := \big(F(X_1, X_2), \big[X_1, X_2\big]\big),$$

for all $v_1, v_2 \in \mathbb{R}^n$ and $X_1, X_2 \in \mathrm{Lie}\,(H)$. Let $\alpha : \mathbb{R}^n \to \mathbb{R}^n \oplus_F \mathrm{Lie}\,(H)$ be the natural injection and $\beta : \mathbb{R}^n \oplus_F \mathrm{Lie}\,(H) \to \mathrm{Lie}\,(H)$ the natural projection. These maps are Lie algebra homomorphisms and Ker β = Im α. Due to Theorems 6 and 7 of Appendix A.1, there exists a unique (up to an isomorphism) connected, simply connected Lie group H_F, such that $\mathrm{Lie}\,(H_F) = \mathbb{R}^n \oplus_F \mathrm{Lie}\,(H)$, and two group homomorphisms $a : \mathbb{R}^n \to H_F$, $b : H_F \to H$ such that the induced Lie algebra homomorphisms $\dot{a}$ and $\dot{b}$ equal α and β, respectively. Moreover, one can prove that a is a homeomorphism from $\mathbb{R}^n$ onto $a(\mathbb{R}^n)$ and $H_F/a(\mathbb{R}^n)$ is isomorphic to H. By a known result (see, for example, Lemma 7.26 of [35]) there exists an analytic map c from H to H_F such that $c(e) = e$ and $b(c(h)) = h$ for all $h \in H$. If we define

$$\tau_F(h_1, h_2) := c(h_1)c(h_2)c(h_1h_2)^{-1}, \qquad h_1, h_2 \in H,$$

then τ_F is (modulo identification) an analytic $\mathbb{R}^n$-multiplier and the function $[F] \to [\tau_F]$ is the isomorphism in question. Since τ_F is analytic, one can easily check that H_F is isomorphic, as a Lie group, to $\mathbb{R}^n \times_{\tau_F} H$, which is a Lie group with respect to the product

$$(v_1, g_1)(v_2, g_2) = (v_1 + v_2 + \tau_F(g_1, g_2), g_1g_2), \qquad v_1, v_2 \in \mathbb{R}^n,\ g_1, g_2 \in H. \quad \square$$

3.3 Universal Central Extension of a Connected Lie Group

Let G be a connected Lie group, G^* its universal covering group and $\delta : G^* \to G$ the covering homomorphism. The kernel of δ, Ker $\delta = \{g^* \in G \mid \delta(g^*) = e\}$, is a closed, discrete, central subgroup of G^*.

Let $H^2(G^*, \mathbb{R})$ be the vector space of the equivalence classes of the $\mathbb{R}$-multipliers of G^*. By Theorem 2 it is finite dimensional. Let $H^2(G^*, \mathbb{R})_\delta$ be the subset of the equivalence classes $[\tau] \in H^2(G^*, \mathbb{R})$ such that

$$\tau(k, g^*) = \tau(g^*, k), \qquad k \in \text{Ker } \delta, \ g^* \in G^*. \tag{3.4}$$

Since Ker δ is central in G^* this relation holds for all $\mathbb{R}$-multipliers of G^* which are equivalent to τ. Hence the set $H^2(G^*, \mathbb{R})_\delta$ is well defined. Moreover, $H^2(G^*, \mathbb{R})_\delta$ is a subspace of $H^2(G^*, \mathbb{R})$. Let N be its dimension.

Let $\tau_1, \ldots, \tau_N$ be some fixed analytic $\mathbb{R}$-multipliers of G^* such that their equivalence classes $[\tau_1], \ldots, [\tau_N]$ form a basis of $H^2(G^*, \mathbb{R})_\delta$. The function $\overline{\tau} : G^* \times G^* \to \mathbb{R}^N$, defined as

$$\overline{\tau}(g_1^*, g_2^*)_i := \tau_i(g_1^*, g_2^*), \qquad g_1^*, g_2^* \in G^*, \ i = 1, \ldots, N,$$

is an analytic $\mathbb{R}^N$-multiplier of G^*. The restriction of $\overline{\tau}$ to Ker $\delta \times$Ker δ is an $\mathbb{R}^N$-multiplier of the discrete group Ker δ, hence it is exact (see Proposition 2, Sect. 4, Chap. 1 of [6]). Without loss of generality one may thus assume that $\overline{\tau}$ is analytic and

$$\overline{\tau}(k_1, k_2) = 0, \qquad k_1, k_2 \in \text{Ker } \delta. \tag{3.5}$$

Definition 13. Let $\overline{G} = \mathbb{R}^N \times G^*$ be the product manifold. Since $\overline{\tau}$ is analytic, $\overline{G}$ is a Lie group with respect to the product

$$(v_1, g_1^*)(v_2, g_2^*) = (v_1 + v_2 + \overline{\tau}(g_1^*, g_2^*), g_1^* g_2^*), \qquad v_1, v_2 \in \mathbb{R}^n, \ g_1^*, g_2^* \in G^*.$$

$\overline{G}$ is the *universal central extension* of G.

The following observation explains why $\overline{G}$ is called a *central extension* of G. Define the map ρ from $\overline{G}$ to G as

$$\rho(v, g^*) := \delta(g^*), \qquad v \in \mathbb{R}^N, \ g^* \in G^*.$$

Clearly, ρ is an analytic surjective group homomorphism and its kernel

$$K = \{(v, k) \in \overline{G} \mid v \in \mathbb{R}^N, \ k \in \text{Ker } \delta\}$$

is a closed subgroup of $\overline{G}$. By definition, $\overline{\tau}(k, g^*) = \overline{\tau}(g^*, k)$ for all $k \in$ Ker δ and $g^* \in G^*$, so that K is central in $\overline{G}$. Hence, $\overline{G}$ is a kind of a generalization of the universal covering group. Moreover, (3.5) implies that K, as a Lie group, is the direct product of $\mathbb{R}^N$ and Ker δ, that is, $K = \mathbb{R}^N \times$ Ker δ.

The next two definitions are essential in order to describe properly the relation between the symmetry actions of G and the unitary representations of $\overline{G}$. We recall the notation $\mathbf{T} = \{zI \mid z \in \mathbb{T}\}$

Definition 14. A representation $U : \overline{G} \to \mathbf{U}$ is *admissible* if it satisfies the condition

$$U_h \in \mathbf{T} \ \text{ for all } h \in K. \tag{3.6}$$

Let U be an admissible representation. Its restriction to K is a character of K. Since $K = \mathbb{R}^N \times \mathrm{Ker}\,\delta$, this character is of the form $U_{(v,k)} = e^{iw\cdot v}\epsilon(k)I$, $v \in \mathbb{R}^N$, $k \in \mathrm{Ker}\,\delta$, for some $w \in \mathbb{R}^N$ and for some character ϵ of Ker δ. We call w the *algebraic charge* of U and ϵ its *topological charge.* The motivation for these names derives from the fact that Ker δ depends only on the topological structure of G, whereas the dimension N is connected with the structure of $H^2(\mathrm{Lie}\,(G), \mathbb{R})$.

Every irreducible representation of $\overline{G}$ is admissible. Indeed, let U be an irreducible representation. Since K is central, U_k, $k \in K$, commutes with each $U_{\overline{g}}$, $\overline{g} \in \overline{G}$, so that, by Schur's lemma, U_k is a multiple of the identity. Since U_k is unitary, it is a phase factor, that is, $U_k \in \mathbf{T}$.

Definition 15. Let U and U' be two unitary representations of $\overline{G}$ acting in $\mathcal{H}$ and $\mathcal{H}'$, respectively. We say that U and U' are *physically equivalent* if there exists a unitary or antiunitary operator $B : \mathcal{H} \to \mathcal{H}'$ and a map $b : \overline{G} \to \mathbb{T}$ such that

$$BU_{\overline{g}} = b(\overline{g})U'_{\overline{g}}B, \qquad \overline{g} \in \overline{G}. \tag{3.7}$$

If U and U' are unitarily equivalent, then they are also physically equivalent, but the converse implication is not true. Moreover, if U is an admissible representation, then every representation that is physically equivalent to U is also admissible.

The following lemma shows that the map b in (3.7) is, in fact, a character of G^*.

Lemma 9. *Let U and U' be two unitary representations of $\overline{G}$. The representations U and U' are physically equivalent if and only if there is a character χ of G^* and a unitary or antiunitary operator B such that (3.7) holds with*

$$b(v, g^*) = \chi(g^*) \qquad v \in \mathbb{R}^N, g^* \in G^*.$$

Proof. Assume that U and U' are physically equivalent and let b and B be such that (3.7) holds. Then, for all $\overline{g} \in \overline{G}$

$$b(\overline{g})I = BU_{\overline{g}}B^{-1}{U'}_{\overline{g}}^{-1}.$$

Fix a unit vector $\varphi \in \mathcal{H}'$. Then $b(\overline{g}) = \langle U_{\overline{g}}^{-1}B^{-1}\varphi, B^{-1}{U'}_{\overline{g}}^{-1}\varphi\rangle$. Since U and U' are continuous in the strong operator topology, the function b is also continuous. Moreover, for all $\overline{g}_1, \overline{g}_2 \in G$

$$\begin{aligned} BU_{\overline{g}_1\overline{g}_2} &= b(\overline{g}_1\overline{g}_2)U'_{\overline{g}_1\overline{g}_2}B \\ BU_{\overline{g}_1}U_{\overline{g}_2} &= b(\overline{g}_1\overline{g}_2)U'_{\overline{g}_1}U'_{\overline{g}_2}B \\ b(\overline{g}_1)b(\overline{g}_2)U'_{\overline{g}_1}U'_{\overline{g}_2}B &= b(\overline{g}_1\overline{g}_2)U'_{\overline{g}_1}U'_{\overline{g}_2}B. \end{aligned}$$

Thus $b(\overline{g}_1)b(\overline{g}_2) = b(\overline{g}_1\overline{g}_2)$. Obviously, $b(e) = 1$, so that b is a character of $\overline{G}$. The restriction of b to $\mathbb{R}^N$ is thus of the form $b(v, e^*) = e^{iw\cdot v}$, $v \in \mathbb{R}^N$, for some $w \in \mathbb{R}^N$. Then, if $g_1^*, g_2^* \in G^*$,

$$\begin{aligned} b((0,g_1^*))b((0,g_2^*)) &= b((\overline{\tau}(g_1^*,g_2^*),e^*))b((0,g_1^*g_2^*)) \\ &= e^{iw\cdot\overline{\tau}(g_1^*,g_2^*)}b((0,g_1^*g_2^*)). \end{aligned}$$

Hence, by Lemma 7, the $\mathbb{R}$-multiplier $w\cdot\overline{\tau}$ is exact, so that $w=0$. If χ is the restriction of b to G^*, then for all $v\in\mathbb{R}^N$ and $g\in G$, $b(v,g^*)=\chi(g^*)$ and χ is a character of G^*. The converse implication is evident. □

We are now prepared state the main result of this section. Let U be an admissible representation of $\overline{G}$ and define, for all $g\in G$,

$$\sigma_g^U := \pi(U_{\overline{g}}), \tag{3.8}$$

where $\overline{g}=(v,g^*)\in\overline{G}$ is such that $\rho(\overline{g})=\delta(g^*)=g$. The following theorem is then obtained.

Theorem 3. *With the above notations, σ^U is a symmetry action of G and the correspondence $[U]\mapsto[\sigma^U]$ between the physical equivalence classes of admissible unitary representations of $\overline{G}$ and the equivalence classes of the symmetry actions of G is a bijection. The representation U of $\overline{G}$ is irreducible if and only if σ^U is an irreducible symmetry action of G.*

Before proving the theorem some comments are due. The above result shows that the equivalence classes of admissible representations of $\overline{G}$ classify quantum systems that are different from each other with respect to the symmetry group G. In particular, the irreducible representations of $\overline{G}$, which are always admissible, describe all the possible systems that are elementary with respect to G.

Consider next the case of a reducible representation of $\overline{G}$. For sake of simplicity, assume that $U=U_1\oplus U_2$, where U_1 and U_2 are irreducible representations, so that the representations U_i are admissible. Let w_i and ϵ_i denote the corresponding algebraic and topological charges. In general, U is not admissible. A simple calculation shows that U is admissible if and only if $w_1=w_2$ and $\epsilon_1=\epsilon_2$. Thus vector states associated with different elementary systems can be superposed into new vector states only if the elementary systems have the same algebraic and topological charges. This fact is at the root of the existence of superselection rules for non-elementary systems.

The relation between the decomposition into irreducible representations and the notion of physical equivalence also requires some special care. One can easily show that, if b is a nontrivial character of $\overline{G}$ which is 1 on K, then, bU_2 is an irreducible representation physically equivalent to U_2. Nevertheless, $U_1\oplus U_2$ and $U_1\oplus b\,U_2$ are physically inequivalent admissible representations. In the same way, if the algebraic charge w of U_1 and U_2 is zero and their topological charge ϵ is such that ϵ^2 extends to a character b of $\overline{G}$, then $U_1\oplus U_2$ and $U_1\oplus bBU_2B^{-1}$, where B is any antiunitary operator, are physically inequivalent admissible representations, even though U_2 and bBU_2B^{-1} are physically equivalent. This kind of phenomenon does not occur if one considers the unitary equivalence instead of the physical equivalence.

We come back to the proof of Theorem 3, which requires some technical lemmas.

We start with stating some properties of $\overline{G}$.

Lemma 10. *Let $\overline{G}$ be the universal central extension of G.*

1. *There is a measurable map $c : G \to \overline{G}$ such that $c(e) = (0, e^*) = \overline{e}$ and $\rho(c(g)) = g$ for all $g \in G$.* (We call such a map a *section* for ρ.)
2. *Given a section c for ρ, define the map $\Gamma_c : G \times G \to K$ as*

$$\Gamma_c(g_1, g_2) := c(g_1)c(g_2)c(g_1 g_2)^{-1}, \qquad g_1, g_2 \in G.$$

   Then Γ_c is a K-multiplier of G and its equivalence class does not depend on the choice of the section c.
3. *Considering $\mathbb{R}^N$ as a subgroup of $K = \mathbb{R}^N \times \mathrm{Ker}\delta$, the K-multiplier $\Gamma_c \circ (\delta \times \delta)$ of G^* is equivalent to $\overline{\tau}$.*
4. *Let χ be a character of K. With the above notations, the map $\mu_\chi : G \times G \to \mathbb{T}$ defined as*

$$\mu_\chi(g_1, g_2) := \chi\left(\Gamma_c(g_1, g_2)\right), \qquad g_1, g_2 \in G,$$

   is a $\mathbb{T}$-multiplier of G and its equivalence class $[\mu_\chi]$ does not depend on the choice of the section c.

Proof.

1. Since ρ is a surjective group homomorphism whose kernel is K and ρ is analytic, G is isomorphic, as a Lie group, to the quotient $\overline{G}/K$. The existence of a section is thus a standard result (see, for example, Theorem 5.11 of [35]).
2. If $g_1, g_2 \in G$, then $\rho(\Gamma_c(g_1, g_2)) = e$, so that $\Gamma_c(g_1, g_2) \in K$. By direct computation one checks that Γ_c is a K-multiplier. Let c' be another section for ρ, then, for all $g \in G$, $c(g) = b(g)c'(g)$ for some measurable function b from G to K. Hence, for all $g_1, g_2 \in G$

$$\Gamma_{c'}(g_1, g_2) = \frac{b(g_1 g_2)}{b(g_1)b(g_2)}\Gamma_c(g_1, g_2).$$

3. Let $i : G^* \to \overline{G}$ be the natural immersion and a the measurable map from G^* to $\overline{G}$ defined as

$$a(g^*) := c(\delta(g^*))i(g^*)^{-1}, \qquad g^* \in G^*.$$

   Since $\rho(a(g^*)) = e$, the map a takes values in K. Then, if $g_1^*, g_2^* \in G^*$,

$$\begin{aligned}\Gamma_c(\delta(g_1^*), \delta(g_2^*)) &= c(\delta(g_1^*))c(\delta(g_1^*))c(\delta(g_1^*)\delta(g_2^*))^{-1} \\ &= a(g_1^*)i(g_1^*)a(g_2^*)i(g_2^*)i(g_1^* g_2^*)^{-1}a(g_1^* g_2^*)^{-1} \\ &= a(g_1^*)a(g_2^*)a(g_1^* g_2^*)^{-1}i(g_1^*)i(g_2^*)i(g_1^* g_2^*)^{-1} \\ &= a(g_1^*)a(g_2^*)a(g_1^* g_2^*)^{-1}(\overline{\tau}(g_1^*, g_2^*), e^*),\end{aligned}$$

   i.e., $\Gamma_c \circ (\delta \times \delta)$ is equivalent to $\overline{\tau}$.
4. This is a simple consequence of the properties of Γ_c given in item 2. □

The following lemma describes the group $H^2(G,\mathbb{T})$ in terms of the characters of K and G^*. This result is important in itself.

Let $\widehat{K}$ be the dual group of K and V the subgroup of those characters of K which extend to characters of $\overline{G}$. By the Lemma 9 the elements of V can be identified with characters of G^*.

Lemma 11. *The mapping $\widehat{K} \ni \chi \mapsto [\mu_\chi] \in H^2(G,\mathbb{T})$ is a surjective homomorphism whose kernel is V.*

Proof. By direct computation one can check that $\chi \mapsto [\mu_\chi]$ is a group homomorphism. To show its surjectivity, we notice that, since the equivalence class $[\mu_\chi]$ does not depend on the specific form of the section c, we can choose for c the particularly simple form

$$c(g) = (0, \tilde{c}(g)) \ , \qquad g \in G,$$

where $\tilde{c} : G \to G^*$ is measurable and satisfies $\tilde{c}(e) = e^*$ and $\delta(\tilde{c}(g)) = g$ for all $g \in G$. With this choice a straightforward calculation shows that

$$\Gamma_c(g_1, g_2) = (\ \overline{\tau}(\tilde{c}(g_1), \tilde{c}(g_2)) - \overline{\tau}(\gamma(g_1, g_2), \tilde{c}(g_1 g_2)) \ , \ \gamma(g_1, g_2)\ ) \ , \tag{3.9}$$

where $g_1, g_2 \in G$ and $\gamma(g_1, g_2) = \tilde{c}(g_1)\tilde{c}(g_2)\tilde{c}(g_1 g_2)^{-1} \in \text{Ker } \delta$. Now let μ be a $\mathbb{T}$-multiplier of G and μ^* the $\mathbb{T}$-multiplier of G^*

$$\mu^*(g_1^*, g_2^*) = \mu(\delta(g_1^*), \delta(g_2^*)), \qquad g_1^*, g_2^* \in G^*.$$

According to Lemma 7,

$$\mu^*(g_1^*, g_2^*) = \frac{a(g_1^* g_2^*)}{a(g_1^*)a(g_2^*)} e^{i\tau(g_1^*, g_2^*)}, \qquad g_1^*, g_2^* \in G^*, \tag{3.10}$$

for some analytic $\mathbb{R}$-multiplier τ of G^* and a measurable function $a : G^* \to \mathbb{T}$. We claim that

$$\tau(k, g^*) = \tau(g^*, k) \qquad k \in \text{Ker } \delta, \ g^* \in G^*. \tag{3.11}$$

In fact, let $k \in \text{Ker } \delta$ and $g^* \in G^*$. Since $\mu^*(k, g^*) = \mu^*(g^*, k) = 1$, then

$$e^{i\tau(k,g^*)} = \frac{a(k)a(g^*)}{a(kg^*)} = \frac{a(k)a(g^*)}{a(g^* k)} = e^{i\tau(g^*,k)}.$$

Hence $\tau(k, g^*) = \tau(g^*, k) + 2\pi n(k, g^*)$ where $n(k, g^*)$ is an integer. By continuity of $\tau(k, \cdot)$ and since G^* is connected, the map $n(\cdot, \cdot)$ depends only on k, and, choosing $g^* = k$, we conclude that $n(k, g^*) = 0$ for all $k \in \text{Ker } \delta$, $g^* \in G^*$.

Due to (3.11), the equivalence class of τ belongs to $H^2(G^*, \mathbb{R})_\delta$ and, by definition of $\overline{\tau}$, there is a $w \in \mathbb{R}^N$ such that, up to equivalence, $\tau = w \cdot \overline{\tau}$. Hence (3.10) becomes

$$\mu^*(g_1^*, g_2^*) = \frac{a(g_1^* g_2^*)}{a(g_1^*)a(g_2^*)} e^{iw\cdot\overline{\tau}(g_1^*, g_2^*)}, \qquad g_1^*, g_2^* \in G^*. \tag{3.12}$$

The previous equality implies that the map $\chi : K \to \mathbb{T}$, with $\chi(v,k) := e^{iw\cdot v}a(k)$, $v \in \mathbb{R}^N$, $k \in \operatorname{Ker}\,\delta$, is, in fact, a character of K. Hence, by statement 4 of Lemma 10, χ defines a $\mathbb{T}$-multiplier μ_χ of G.

We will show that μ_χ is equivalent to μ. In fact, using (3.9), one has

$$\begin{aligned}\mu_\chi(g_1, g_2) &= \chi(\Gamma_c(g_1, g_2)) \\ &= e^{iw\cdot(\,\overline{\tau}(\tilde{c}(g_1), \tilde{c}(g_2)) - \overline{\tau}(\gamma(g_1,g_2), \tilde{c}(g_1 g_2))\,)} a(\gamma(g_1, g_2)).\end{aligned}$$

Using twice (3.12) we obtain

$$\begin{aligned} e^{iw\cdot\overline{\tau}(\tilde{c}(g_1), \tilde{c}(g_2))} &= \frac{a(\tilde{c}(g_1))a(\tilde{c}(g_2))}{a(\tilde{c}(g_1)\tilde{c}(g_2))}\mu(g_1, g_2) \\ e^{-iw\cdot\overline{\tau}(\gamma(g_1,g_2), \tilde{c}(g_1 g_2))} &= \frac{a(\tilde{c}(g_1)\tilde{c}(g_2))}{a(\gamma(g_1, g_2))a(\tilde{c}(g_1 g_2))}\end{aligned}$$

so that

$$\mu_\chi(g_1, g_2) = \frac{a(\tilde{c}(g_1))a(\tilde{c}(g_2))}{a(\tilde{c}(g_1 g_2))}\mu(g_1, g_2),$$

which shows the equivalence of μ and μ_χ.

Suppose now that χ is a character of K that extends to a character of $\overline{G}$ (still denoted by χ). Then

$$\begin{aligned}\mu_\chi(g_1, g_2) &= \chi(c(g_1)c(g_2)c(g_1 g_2)^{-1}) \\ &= \chi(c(g_1))\chi(c(g_2))\chi(c(g_1 g_2)^{-1}),\end{aligned}$$

showing that μ_χ is exact. Conversely, assume that

$$\mu_\chi(g_1, g_2) = \frac{a(g_1 g_2)}{a(g_1)a(g_2)}$$

for some measurable function $a : G \to \mathbb{T}$. Observe that, for all $\overline{g} \in \overline{G}$, $\overline{g}c(\rho(\overline{g}))^{-1} \in K$ and define $\chi' : \overline{G} \to \mathbb{T}$ as

$$\chi'(\overline{g}) = \chi(hc(\rho(\overline{g}))^{-1})a(\rho(\overline{g}))^{-1}, \qquad \overline{g} \in \overline{G}.$$

Then χ' is a character of $\overline{G}$. Indeed, χ' is measurable, and if $\overline{g}_1, \overline{g}_2 \in \overline{G}$,

$$\begin{aligned}\chi'(\overline{g}_1)\chi'(\overline{g}_2) &= \frac{\chi(\overline{g}_1 c(\rho(\overline{g}_1))^{-1}\overline{g}_2 c(\rho(\overline{g}_2))^{-1})}{a(\rho(\overline{g}_1))a(\rho(\overline{g}_2))} \\ &= \frac{\chi\left(\overline{g}_1\overline{g}_2 c(\rho(\overline{g}_2))^{-1}c(\rho(\overline{g}_1))^{-1}\right)\mu_\chi(g_1, g_2)}{a(\rho(\overline{g}_1\overline{g}_2))} \\ &= \frac{\chi\left(\overline{g}_1\overline{g}_2 c(\rho(\overline{g}_2))^{-1}c(\rho(\overline{g}_1))^{-1}c(\rho(\overline{g}_1))c(\rho(\overline{g}_2))c(\rho(\overline{g}_1\overline{g}_2))^{-1}\right)}{a(\rho(\overline{g}_1\overline{g}_2))} \\ &= \chi\left(\overline{g}_1\overline{g}_2 c(\rho(\overline{g}_1\overline{g}_2))^{-1}\right)a(\rho(\overline{g}_1\overline{g}_2))^{-1} \\ &= \chi'(\overline{g}_1\overline{g}_2).\end{aligned}$$

Moreover, since $a(e) = 1$, $\chi'(k) = \chi(k)$ for all $k \in K$.

Hence, $H^2(G, \mathbb{T})$ is isomorphic, as an abstract group, to the quotient group $\hat{K}/V$ and this concludes the proof. □

If the subgroup V of $\widehat{K}$ is closed one may give a better description of $H^2(G, \mathbb{T})$. Define

$$K_0 := \{(v, k) \in K \mid b(k) = 1 \text{ for any character } b \text{ of } G^*\}.$$

Then K_0 is a commutative closed subgroup of K. Since V is closed, a standard result on commutative locally compact groups (see, for example, Theorem 4.39 of [14]) shows that $\widehat{K}/V$ is isomorphic to the dual group $\widehat{K_0}$ of K_0. In particular, any element $\chi \in \widehat{K_0}$ extends to an element $\widehat{\chi} \in \widehat{K}$ and $\widehat{\chi}$ is uniquely defined by χ, up to an element of V. Let μ_χ be the $\mathbb{T}$-multiplier of G defined by

$$\mu_\chi(g_1, g_2) = \widehat{\chi}(\Gamma_c(g_1, g_2)), \qquad g_1, g_2 \in G,$$

where Γ_c is defined in Lemma 10. As a consequence of Lemma 11, the equivalence class $[\mu_\chi]$ depends only on χ and not on the particular extension chosen.

Corollary 5. *If V is closed, the map $\widehat{K_0} \ni \chi \mapsto [\mu_\chi] \in H^2(G, \mathbb{T})$ is a group isomorphism.*

We are now ready to prove the main theorem of this section.

Proof (Proof of Theorem 3). In the following we fix a section $c : G \to \overline{G}$ for the function $\rho : \overline{G} \to G$ and a section $s : \Sigma_0 \to \mathbf{U}$ for the canonical projection $\pi : \mathbf{U} \to \Sigma_0$. Due to the admissibility condition (3.6), if $h_1, h_2 \in \overline{G}$ are such that $\rho(h_1) = \rho(h_2) = g$, then $\pi(U_{h_1}) = \pi(U_{h_2})$, showing that σ^U_g is well-defined. In particular, we have

$$\sigma^U_g = \pi(U_{c(g)}) \,, \qquad g \in G.$$

First we show that $g \mapsto \sigma^U_g$ is a symmetry action of G. Indeed, if $g_1, g_2 \in G$, then

$$\begin{aligned} \sigma^U_{g_1}\sigma^U_{g_2} &= \pi(U_{c(g_1)})\pi(U_{c(g_2)}) \\ &= \pi(U_{c(g_1)}U_{c(g_2)}) \\ &= \pi(U_{c(g_1)c(g_2)c(g_1g_2)^{-1}})\pi(U_{c(g_1g_2)}) \\ &= \pi(U_{c(g_1g_2)}) = \sigma^U_{g_1g_2}, \end{aligned}$$

where we used the fact that $c(g_1)c(g_2)c(g_1g_2)^{-1} \in K$ as well as the admissibility of U. Since c is measurable, σ^U is also measurable. Also, $\sigma^U_e = I$, so that σ^U is a symmetry action of G.

Let U and U' be two physically equivalent admissible representations of $\overline{G}$ acting on $\mathcal{H}$ and $\mathcal{H}'$, respectively. The corresponding symmetry actions σ^U

and $\sigma^{U'}$ are also equivalent. Indeed, in this case $BU_h = b(h)U'_h B$, $h \in \overline{G}$, for some unitary or antiunitary operator $B : \mathcal{H} \to \mathcal{H}'$ and a character $b : \overline{G} \to \mathbb{T}$. If β denotes the equivalence class $[B] \in \Sigma(\mathcal{H}, \mathcal{H}')$, then $\beta\sigma_g^U = \sigma_g^{U'}\beta$ for all $g \in G$, which is just to say that σ^U and $\sigma^{U'}$ are equivalent. This shows that the map $[U] \mapsto [\sigma^U]$ is well defined.

We now show its surjectivity. Let σ be a symmetry action of G and define $\mu : G \times G \to \mathbf{U}$ as

$$\mu(g_1, g_2) := s(\sigma_{g_1})s(\sigma_{g_2})s(\sigma_{g_1 g_2})^{-1}, \qquad g_1, g_2 \in G.$$

Since $\pi(\mu(g_1, g_2)) = I$, $\mu(g_1, g_2) \in \mathbf{T}$. Moreover, μ is measurable and by a direct computation one confirms that μ is, in fact, a $\mathbb{T}$-multiplier of G. By Lemma 11, there is a character χ of K and a measurable function $a : G \to \mathbb{T}$ such that

$$\mu(g_1, g_2) = \frac{a(g_1 g_2)}{a(g_1)a(g_2)}\mu_\chi(g_1, g_2)\,, \qquad g_1, g_2 \in G. \tag{3.13}$$

Define a map $U^\sigma : \overline{G} \to \mathbf{U}$ as

$$U_h^\sigma := \chi(hc(\rho(h))^{-1})a(\rho(h))s(\sigma_{\rho(h)}), \qquad h \in \overline{G}.$$

Then U^σ is a representation of $\overline{G}$. Indeed, as a composition of measurable maps U^σ is measurable. Since $a(e) = 1$ and $s(I) = I$, $U^\sigma_{(0,e^*)} = I$. Finally, for any $h_1, h_2 \in \overline{G}$,

$$\begin{aligned}
U_{h_1}^\sigma U_{h_2}^\sigma &= \chi(h_1 c(\rho(h_1))^{-1} h_2 c(\rho(h_2))^{-1})a(\rho(h_1))a(\rho(h_2))s(\sigma_{\rho(h_1)})s(\sigma_{\rho(h_2)}) \\
&= \chi\left(h_1 h_2 c(\rho(h_2))^{-1} c(\rho(h_1))^{-1}\right) \\
&\qquad a(\rho(h_1))a(\rho(h_2))\mu(g_1, g_2)s(\sigma_{\rho(h_1 h_2)}) \\
&= \chi\left(h_1 h_2 c(\rho(h_2))^{-1} c(\rho(h_1))^{-1}\right)\chi\left(c(\rho(h_1))c(\rho(h_2))c(\rho(h_1 h_2))^{-1}\right) \\
&\qquad a(\rho(h_1 h_2))s(\sigma_{\rho(h_1 h_2)}) \\
&= \chi\left(h_1 h_2 c(\rho(h_1 h_2))^{-1}\right)a(\rho(h_1 h_2))s(\sigma_{\rho(h_1 h_2)}) \\
&= U_{h_1 h_2}^\sigma .
\end{aligned}$$

Since $\pi \circ s = id_{\Sigma_0}$ and $\rho \circ c = id_G$, one readily verifies that $\sigma^{U^\sigma} = \sigma$, proving the surjectivity of the map $[U] \mapsto [\sigma^U]$.

Assume next that σ^U and $\sigma^{U'}$ are equivalent symmetry actions, and let $\beta \in \Sigma(\mathcal{H}, \mathcal{H}')$ be such that $\pi(U'_{c(g)})\beta = \beta\pi(U_{c(g)})$, for all $g \in G$. We may thus conclude that for some unitary or antiunitary operator $B \in \beta$ and for some measurable map $b : G \to \mathbb{T}$, $U'_{c(g)} = b(g)BU_{c(g)}B^{-1}$. Let $h \in \overline{G}$, $g = \rho(h)$, and $k = hc(g)^{-1}$, then $k \in K$ and

$$\begin{aligned}
U'_h = U'_k U'_{c(g)} &= U'_k b(c(g))BU_{c(g)}B^{-1} \\
= U'_k b(c(g))BU_{k^{-1}}U_h B^{-1} &= \hat{b}(h)BU_h B^{-1},
\end{aligned}$$

taking into account that, due to (3.6), U'_k and $U_{k^{-1}}$ are phase factors that we have collected in $\hat{b}$. This shows that U and U' are physically equivalent representations of $\overline{G}$, proving the injectivity of the map $[U] \mapsto [\sigma^U]$.

To conclude, we prove the statement about irreducibility. Given, as above, $h = c(g)k \in \overline{G}$ with $g \in G$ and $k \in K$, and two vectors states $P_1 = P[\phi_1]$ and $P_2 = P[\phi_2]$, with $\phi_1, \phi_2 \in \mathcal{H}$, one has

$$\begin{aligned} |\langle \phi_1, U_h \phi_2 \rangle|^2 &= |\langle \phi_1, U_{c(g)} U_k \phi_2 \rangle|^2 \\ &= |U_k \langle \phi_1, U_{c(g)} U_k \phi_2 \rangle|^2 \\ &= \mathrm{tr}\big[P_1 \sigma_g^U (P_2)\big], \end{aligned} \tag{3.14}$$

where U_k is a phase factor since U is admissible.

We now assume that U is irreducible and we prove that σ^U is also irreducible. Let $P_1 = P[\phi_1]$ and $P_2 = P[\phi_2]$ be two vector states with $\phi_1, \phi_2 \in \mathcal{H}$. Due to the irreducibility of U,

$$\phi_2 \in \overline{\mathrm{span}}\{U_h \phi_1 \ : \ h \in \overline{G}\}$$

so that there is $h \in \overline{G}$ such that $\langle \phi_1, U_h \phi_2 \rangle \neq 0$. By (3.14), it follows that $\mathrm{tr}\big[P_1 \sigma_g^U (P_2)\big] \neq 0$, where $g = \rho(h)$, that is, σ^U is irreducible. The converse statement can be proved in a similar way so that $U : \overline{G} \to \mathbf{U}$ is irreducible if and only if $\sigma^U : G \to \Sigma$ is irreducible. □

Let U be an admissible representation of $\overline{G}$ and w and ϵ its algebraic and topological charges. One can easily check that the map $G^* \ni g^* \mapsto U_{(0,g^*)} \in \mathbf{U}$ is a projective representation of G^* with the $\mathbb{T}$-multiplier $\mu^*(g_1^*, g_2^*) = e^{iw\cdot\bar{\tau}(g_1^*, g_2^*)}$. Moreover, if $c : G \to \overline{G}$ is a section for ρ, then the map $G \ni g \mapsto U_{c(g)} \in \mathbf{U}$ is a projective representation of G and its $\mathbb{T}$-multiplier is μ_χ where $\chi(v, k) = e^{iw\cdot v}\epsilon(k)$ and μ_χ is defined in item 4 of Lemma 10. As a consequence of statement 3 of the same lemma, μ^* and $\mu_\chi \circ (\delta \times \delta)$ are equivalent. Nevertheless, even if μ^* is exact, μ_χ could be *non-exact*.

3.4 The Physical Equivalence for Semidirect Products

According to Theorem 3, the irreducible inequivalent symmetry actions of a group G are completely described by the irreducible physically inequivalent representations of its universal central extension $\overline{G}$. In the examples considered in this monograph, the universal central extension is a regular semidirect product with a commutative normal subgroup, so that any irreducible representation is unitarily equivalent to some induced one [26]. In this way, the problem of characterizing physically inequivalent irreducible representations is reduced to the analogous problem for the induced representation. The present section describes the solution in terms of properties of the orbits in the dual space and of the inducing representations.

Let $\overline{G} = A \times' H$ be a Lie group with A a commutative normal closed subgroup and H a closed subgroup. In this section we denote the elements of $\overline{G}$ as $g = (a, h)$. We denote by $\hat{A}$ the dual group of A and by $(g, \cdot) \mapsto g[\cdot]$ both the inner action of $\overline{G}$ on A and the dual action of $\overline{G}$ on $\hat{A}$. If $x \in \hat{A}$, let $\overline{G}_x := \{g \in \overline{G} \,|\, g[x] = x\}$ be the stability subgroup of $\overline{G}$ at x and $\overline{G}[x] := \{g[x] \,|\, g \in \overline{G}\}$ the corresponding orbit. We assume that each orbit in $\hat{A}$ is locally closed (i.e., the semidirect product is regular) and, to simplify the exposition, that it has a $\overline{G}$-invariant σ-finite measure.

Moreover, given $x \in \hat{A}$ and a representation D of $\overline{G}_x \cap H$ acting in a Hilbert space $\mathcal{K}$, we denote by $U = \mathrm{Ind}_{\overline{G}_x}^{\overline{G}}(xD)$ the representation of $\overline{G}$ unitarily induced by the representation xD of $\overline{G}_x$,

$$(xD)_{ah} = x_a D_h, \qquad a \in A,\ h \in \overline{G}_x \cap H.$$

Explicitly, let ν be a $\overline{G}$-invariant σ-finite measure on $\overline{G}[x]$ and c a measurable map from $\overline{G}[x]$ to $\overline{G}$ such that $c(x) = e$ and $c(y)[x] = y$ for all $y \in \overline{G}[x]$ (we call such a map a *section* for $\overline{G}[x]$). Then U acts on the Hilbert space $L^2(\overline{G}[x], \nu, \mathcal{K})$ as

$$(U_g f)(y) = (xD)_{(c(y)^{-1} g c(g^{-1}[y]))} f(g^{-1}[y]),$$

where $y \in \overline{G}[x]$, $f \in L^2(\overline{G}[x], \nu, \mathcal{K})$, and $g \in \overline{G}$.

We shall now classify all the equivalence classes (with respect to the notion of physical equivalence) of irreducible representations of $\overline{G}$ in the case of regular semidirect products.

Let $\hat{A}_s$ be the set of singleton $\overline{G}$-orbits in $\hat{A}$, i.e.,

$$\hat{A}_s = \{y \in \hat{A} \ :\ g[y] = y, \quad g \in \overline{G}\}.$$

Define for all $x \in \hat{A}$ the *orbit class*

$$\widetilde{\mathcal{O}}_x := \{y g[x^{\epsilon}] \ :\ y \in \hat{A}_s, \quad g \in \overline{G},\ \epsilon = \pm 1\}.$$

Obviously, for all $x' \in \widetilde{\mathcal{O}}_x$, $\overline{G}[x'] \subset \widetilde{\mathcal{O}}_x$ and $\widetilde{\mathcal{O}}_x = \widetilde{\mathcal{O}}_{x'}$, so that one may choose a family $\{x_i\}_{i \in I}$ of elements in $\hat{A}$ such that $\hat{A}$ is the disjoint union of the sets $\widetilde{\mathcal{O}}_{x_i}$.

Theorem 4. *Let* $\overline{G} = A \times' H$ *be a regular semidirect product and* $\hat{A} = \cup_{i \in I} \widetilde{\mathcal{O}}_{x_i}$, $\widetilde{\mathcal{O}}_{x_i} \cap \widetilde{\mathcal{O}}_{x_j} = \emptyset$, *for all* $i \neq j$.

1. *Every irreducible representation of* $\overline{G}$ *is physically equivalent to one of the form* $\mathrm{Ind}_{\overline{G}_{x_i}}^{\overline{G}}(x_i D)$ *for some index* i *and some irreducible representation* D *of* $\overline{G}_{x_i} \cap H$.
2. *If* $i \neq j$ *and* D, D' *are two representations of* $\overline{G}_{x_i} \cap H$ *and* $\overline{G}_{x_j} \cap H$, *respectively, then* $\mathrm{Ind}_{\overline{G}_{x_i}}^{\overline{G}}(x_i D)$ *and* $\mathrm{Ind}_{\overline{G}_{x_j}}^{\overline{G}}(x_j D')$ *are physically inequivalent.*

3. Let $x \in \hat{A}$ and D, D' be two representations of $\overline{G}_x \cap H$. Then $\mathrm{Ind}_{\overline{G}_x}^{\overline{G}}(xD)$ and $\mathrm{Ind}_{\overline{G}_x}^{\overline{G}}(xD')$ are physically equivalent if and only if one of the following two conditions is satisfied:
a) there exist $y \in \hat{A}_s$, a character χ of H, and a unitary operator M such that

$$\begin{aligned} \overline{G}[x] &= y\overline{G}[x], \\ D'_{hsh^{-1}} &= \chi_s M D_s M^{-1}, \qquad s \in \overline{G}_x \cap H, \end{aligned}$$

where $h \in H$ is such that $x = yh[x]$;
b) there exist $y \in \hat{A}_s$, a character χ of H, and an antiunitary operator M such that

$$\begin{aligned} \overline{G}[x] &= y\overline{G}[x]^{-1}, \\ D'_{hsh^{-1}} &= \chi_s M D_s M^{-1}, \qquad s \in \overline{G}_x \cap H, \end{aligned}$$

where $h \in H$ is such that $x = yh[x^{-1}]$.

Motivated by the above theorem, if U is a representation of $\overline{G}$ physically equivalent to some induced representation $\mathrm{Ind}_{\overline{G}_x}^{\overline{G}}(xD)$ we say, with a slight abuse of terminology, that U *lives* on the orbit class $\widetilde{\mathcal{O}}_x$.

The proof of the theorem is based on the following lemma.

Lemma 12. *Let $x, x' \in \hat{A}$. Let D be a representation of $\overline{G}_x \cap H$ acting in $\mathcal{K}$ and D' a representation of $\overline{G}_{x'} \cap H$ acting in $\mathcal{K}'$. The induced representations* $\mathrm{Ind}_{\overline{G}_x}^{\overline{G}}(xD)$ *and* $\mathrm{Ind}_{\overline{G}_{x'}}^{\overline{G}}(x'D')$ *are physically equivalent if and only if there exist an element $h \in \overline{G}$, a character $\widetilde{\chi}$ of $\overline{G}$ and a unitary or antiunitary operator M from $\mathcal{K}$ onto $\mathcal{K}'$ such that*

1. $\overline{G}_{x'} = h\overline{G}_x h^{-1}$;
2. $(x'D')_{hgh^{-1}} = \widetilde{\chi}_g M (xD)_g M^{-1}$ for all $g \in \overline{G}_x$.

Moreover, every character of $\overline{G}$ is of the form

$$(a,h) \mapsto \hat{\chi}_a \chi_h\,, \qquad a \in A,\ h \in H$$

where $\hat{\chi} \in \hat{A}_s$ and χ is a character of H.

Proof. First we prove the statement on the characters of $\overline{G}$. If $\widetilde{\chi}$ is a character of $\overline{G}$, let $\hat{\chi}$ and χ be its restrictions to A and H, respectively. Then χ is a character of H and, by definition of dual action, $\hat{\chi} \in \hat{A}_s$. The proof of the converse implication is similar.

We now turn to the first statement. To simplify the notations, denote $U = \mathrm{Ind}_{\overline{G}_x}^{\overline{G}}(xD)$ and $U' = \mathrm{Ind}_{\overline{G}_{x'}}^{\overline{G}}(x'D')$. The representations U and U' are physically equivalent if and only if there exist a character $\widetilde{\chi}$ of $\overline{G}$ and a unitary or antiunitary operator B such that

$$U' = \widetilde{\chi} B^{-1} U B.$$

As a first step we define in terms of U and $\widetilde{\chi}$ two induced representations U^+ and U^- of $\overline{G}$ such that

$$U^{\pm} = \widetilde{\chi} W_{\pm}^{-1} U W_{\pm},$$

where W_+ [or W_-] is unitary [or antiunitary]. In particular, U^+ and U^- are physically equivalent to U.

By the previous result $\widetilde{\chi} = \hat{\chi}\chi$, where $\hat{\chi} \in \hat{A}_s$ and χ is a character of H. Define the maps ψ_+ and ψ_- from $\hat{A}$ onto $\hat{A}$ as $\psi_{\pm}(x) := \hat{\chi} x^{\pm 1}$. The maps $\psi_{\pm}$ are measurable isomorphisms that commute with the action of $\overline{G}$, so that, for all $x \in \hat{A}$, $\psi_{\pm}$ maps the orbit $\overline{G}[x]$ onto the orbit $\overline{G}[\psi_{\pm}(x)]$ and one has $\overline{G}_x = \overline{G}_{\psi_{\pm}(x)}$. If ν is an invariant measure on $\overline{G}[x]$, the image measure $\nu^{\pm}$ with respect to $\psi_{\pm}$ is an invariant measure on $\overline{G}[\psi_{\pm}(x)]$ and if c is a section for the orbit $\overline{G}[x]$, then $c^{\pm} = c \circ \psi_{\pm}^{-1}$ is a section for the action of $\overline{G}$ on the orbit $\overline{G}[\psi_{\pm}(x)]$.

Fix a unitary operator L_+ and an antiunitary operator L_- on $\mathcal{K}$. Consider the representations of $\overline{G}_x$,

$$g \mapsto \widetilde{\chi}_g L_{\pm} (xD)_g L_{\pm}^{-1},$$

and observe that their restriction to A are exactly the elements $x_{\pm}$. Since $\overline{G}_{x_{\pm}} = \overline{G}_x$ we can define the induced representations of $\overline{G}$,

$$U^{\pm} := \mathrm{Ind}_{\overline{G}_{x_{\pm}}}^{\overline{G}} (\widetilde{\chi} L_{\pm} x D L_1^{-1} \pm),$$

acting in $L^2(\overline{G}[x_{\pm}], \nu^{\pm}, \mathcal{K})$.

Moreover, define the operators $W_{\pm}$ from $L^2(\overline{G}[x_{\pm}], \nu^{\pm}, \mathcal{K})$ onto $L^2(\overline{G}[x], \nu, \mathcal{K})$

$$(W_{\pm} f)(y) = \widetilde{\chi}_{c(y)}^{\pm 1} L_{\pm}^{-1} f(\psi_{\pm}(y)), \qquad y \in \overline{G}[x].$$

It is easy to show that W_+ [or W_-] is unitary [or antiunitary].

We have

$$U^{\pm} = \widetilde{\chi} W_{\pm}^{-1} U W_{\pm}.$$

In fact, let $g \in \overline{G}$, $f \in L^2(\overline{G}[x_{\pm}], \nu^{\pm}, \mathcal{K})$, and $y \in \overline{G}[x_{\pm}]$

$$\begin{aligned}
\widetilde{\chi}_g \left(W_{\pm}^{-1} U_g W_{\pm} f\right)(y) &= \widetilde{\chi}_g \widetilde{\chi}_{c^{\pm}(y)}^{-1} L_{\pm} \left(U_g W_{\pm} f\right)\left(\psi_{\pm}^{-1}(y)\right) \\
&= \widetilde{\chi}_g \widetilde{\chi}_{c^{\pm}(y)}^{-1} L_{\pm} (xD)_{\gamma^{\pm}(g,y)} (W_{\pm} f)(g^{-1}[\psi_{\pm}^{-1}(y)]) \\
&= \widetilde{\chi}_g \widetilde{\chi}_{c^{\pm}(y)}^{-1} L_{\pm} (xD)_{\gamma^{\pm}(g,y)} \widetilde{\chi}_{c^{\pm}(g^{-1}[y])}^{\pm 1} L_{\pm}^{-1} f(g^{-1}[y]) \\
&= (\widetilde{\chi} L_{\pm} x D L_{\pm}^{-1})_{\gamma^{\pm}(g,y)} f(g^{-1}[y]) \\
&= (U_g^{\pm} f)(y).
\end{aligned}$$

where $\gamma^{\pm}(g,y) = c^{\pm}(y)^{-1} g c^{\pm}(g^{-1}[y]) = c(\psi_{\pm}^{-1}(y))^{-1} g c(g^{-1}[\psi_{\pm}^{-1}(y)])$.

To conclude the proof of the lemma, observe first that there always exists a unitary operator V such that either $B = W_+V$ or $B = W_-V$, according to the fact that B is unitary or antiunitary. Hence U and U' are physically equivalent if and only if U' is unitarily equivalent either to U^+ or to U^-. Due to a theorem of Mackey (see, for example, Theorem 6.42 of [14]), this is possible if and only if there exist $h \in \overline{G}$ such that $\overline{G}_{x'} = h\overline{G}_x h^{-1}$ and a unitary or antiunitary operator M (depending on the fact that B is unitary or antiunitary) such that $(x'D')_{hgh^{-1}} = \widetilde{\chi}_g M(xD)_g M^{-1}$ for all $g \in \overline{G}_x$. □

Proof (Proof of Theorem 4).

1. Since the semidirect product is regular, a theorem of Mackey (see, for example, Theorem 6.42 of [14]) implies that each irreducible unitary representation of $\overline{G}$ is unitarily (hence physically) equivalent to one of the form $\mathrm{Ind}_{\overline{G}_x}^{\overline{G}}(xD')$ for some $x \in \hat{A}$ and some irreducible representation D of $\overline{G}_x \cap H$. There is an index i such that $x \in \widetilde{\mathcal{O}}_{x_i}$ and, by definition of orbit class, there exist $y \in \hat{A}_s$ and $h \in \overline{G}$ such that $x = yh[x_i^\epsilon]$ where $\epsilon = \pm 1$. Hence $G_x = hG_{x_i}h^{-1}$ and we can define a representation D of $G_{x_i} \cap H$ either as
$$D_g = D'_{h^{-1}gh}, \qquad g \in G_{x_i} \cap H,$$
if $\epsilon = 1$, or as
$$D_g = MD'_{h^{-1}gh}M^{-1}, \qquad g \in G_{x_i} \cap H,$$
if $\epsilon = -1$, where M is a fixed antiunitary operator. Then, by Lemma 12, $\mathrm{Ind}_{\overline{G}_x}^{\overline{G}}(xD')$ is physically equivalent to $\mathrm{Ind}_{\overline{G}_{x_i}}^{\overline{G}}(x_iD)$.
2. If $\mathrm{Ind}_{\overline{G}_{x_i}}^{\overline{G}}(x_iD)$ and $\mathrm{Ind}_{\overline{G}_{x_j}}^{\overline{G}}(x_jD')$ are physically equivalent, condition 2 of Lemma 12 with the choice $g = a \in A$ implies that $x_j = yh[x_i^\epsilon]$ for some $y \in \hat{A}_s$ and $\epsilon = \pm 1$, so that, by definition of x_i, $i = j$.
3. Apply Lemma 12 with $x = x'$, taking into account the form of the characters of $\overline{G}$. □

We observe that if D' is unitarily equivalent to D, the conditions (a) of item 3 of Theorem 4 are satisfied with $y = 1$, $\widetilde{\chi} = 1$, and $h = e$ and this is exactly the case of unitary equivalence of the induced representations. However, in general, there are other possibilities apart from the unitary equivalence. There are even situations in which both conditions (a) and (b) hold.

3.5 An Example: The Temporal Evolution of a Closed System

As a first simple illustration of the general theory developed so far, consider the additive group of the real line $\mathbb{R}$. It is a connected and simply connected

Lie group, so that, in particular, its covering group $\mathbb{R}^*$ is $\mathbb{R}$ itself, and the symmetry actions of $\mathbb{R}$ on $\mathcal{H}$ take values in Σ_0, $\mathbb{R} \ni t \mapsto \sigma_t \in \Sigma_0$. The Lie algebra $\mathrm{Lie}\,(\mathbb{R})$ of the additive group $\mathbb{R}$ can be identified with the vector space $\mathbb{R}$, with the Lie product $[x, y] = 0$, $x, y \in \mathbb{R}$, and with the exponential map $\mathrm{Lie}\,(\mathbb{R}) \ni x \mapsto \exp(x) = x \in \mathbb{R}$ being the identity. Any bilinear function F on $\mathbb{R} \times \mathbb{R}$ is of the form $F(x, y) = \lambda xy$ for some $\lambda \in \mathbb{R}$, so that there is no skew symmetric bilinear forms in the present case, and, in particular, the vector space of the closed forms $H^2(\mathrm{Lie}\,(\mathbb{R}), \mathbb{R})$ contains only the zero vector. The central universal extension of $\mathbb{R}$ is thus $\mathbb{R}$, $\overline{\mathbb{R}} = \mathbb{R}^* = \mathbb{R}$. According to the Stone theorem, any (strongly continuous) unitary representation $\mathbb{R} \ni t \mapsto U_t \in \mathbf{U}$ is of the form $U_t = e^{itH}$, $t \in \mathbb{R}$, where H is a selfadjoint operator acting in $\mathcal{H}$. Any symmetry action $\sigma : \mathbb{R} \to \Sigma_0$ is now of the form $\sigma = \sigma^U$ for some unitary representation $U : t \mapsto U_t = e^{itH}$, and two symmetry actions σ_1 and σ_2 are equivalent if and only if the representations U_1 and U_2 differ by a character, that is, $H_1 = H_2 + aI$ for some real number a. The temporal evolution of a closed system is a particular instance of the symmetry actions $\mathbb{R} \to \Sigma_0$, and we may conclude that two systems with Hamiltonians H_1 and H_2 which differ only by a constant aI do have the same evolutions.

4 The Galilei Groups

In this chapter we describe the Galilei group and its universal central extension both in $3+1$ and in $2+1$ dimensions.

4.1 The $3+1$ Dimensional Case

In this section we use the vector notation $\boldsymbol{x}$ for the elements of $\mathbb{R}^3$.

Let $\mathcal{V} := (\mathbb{R}^3, +)$ be the three dimensional real vector group, the group of velocity transformations, and let $SO(3)$ be the classical Lie group of the special orthogonal 3×3 real matrices, the rotation group in $\mathbb{R}^3$. For any $(\boldsymbol{v}, R) \in \mathcal{V} \times SO(3)$ we put, following the notation of the semidirect product in A.1,

$$f(\boldsymbol{v}, R) \equiv R[\boldsymbol{v}] := R\boldsymbol{v}.$$

This defines an analytic action of $SO(3)$ on $\mathcal{V}$ and the semidirect product defined by this action is the *homogeneous Galilei group*

$$G_o := \mathcal{V} \times' SO(3).$$

According to the definition of Lie subgroup in A.1 both $\mathcal{V}$ and $SO(3)$ are closed Lie subgroups of G_o, and $\mathcal{V}$ is a normal subgroup of G_o. In addition, $\mathcal{V}$ is Abelian, and $SO(3)$ is compact and connected but not simply connected [38].

Let $\mathcal{T}_s := (\mathbb{R}^3, +)$ be the three dimensional real vector group, the group of space translations, $\mathcal{T}_t := (\mathbb{R}, +)$ the one dimensional real vector group, the group of time translations, and let $\mathcal{T} := \mathcal{T}_s \times \mathcal{T}_t$ denote the four dimensional real vector group of space-time translations. We denote its elements by $(\boldsymbol{a}, b)$. Consider the action of G_o on $\mathcal{T}$ defined as

$$(\boldsymbol{v}, R)[(\boldsymbol{a}, b)] := (R\boldsymbol{a} + b\boldsymbol{v}, b)$$

for all $(\boldsymbol{v}, R) \in G_o, (\boldsymbol{a}, b) \in \mathcal{T}$. This is an analytic action and it defines the semidirect product, *the Galilei group*

$$G := \mathcal{T} \times' G_o.$$

As above, $\mathcal{T}$ and G_o are closed Lie subgroups of G, and $\mathcal{T}$ is normal and Abelian.

G. Cassinelli, E. De Vito, P.J. Lahti, and A. Levrero, *The Theory of Symmetry Actions in Quantum Mechanics*, Lect. Notes Phys. **654**, pp. 49–59
http://www.springerlink.com/

For any $g \in G$ we write $g = (\boldsymbol{a}, b, \boldsymbol{v}, R)$ and we call these the Galilei transformations. The identity element of G is $e = (\boldsymbol{0}, 0, \boldsymbol{0}, I)$ and the inverse of an element g is given as

$$g^{-1} = (\boldsymbol{a}, b, \boldsymbol{v}, R)^{-1} = (-R^{-1}(\boldsymbol{a} - b\boldsymbol{v}), -b, -R^{-1}\boldsymbol{v}, R^{-1}).$$

The product of two transformations g, g' obtains the form

$$gg' = (\boldsymbol{a}, b, \boldsymbol{v}, R)(\boldsymbol{a}', b', \boldsymbol{v}', R') = (R\boldsymbol{a}' + b'\boldsymbol{v} + \boldsymbol{a}, b + b', R\boldsymbol{v}' + \boldsymbol{v}, RR').$$

4.1.1 Physical Interpretation

The Galilei group G acts as a Lie transformation group on $\mathbb{R}^4$:

$$g[(\boldsymbol{x}, t)] := (R\boldsymbol{x} + \boldsymbol{v}t + \boldsymbol{a}, b + t)$$

for any $g \in G$ and for all $(\boldsymbol{x}, t) \in \mathbb{R}^4$. This action allows one to identify G as the group of transformations of the coordinates of the Newtonian space-time $\mathbb{R}^4 = \mathbb{R}^3 \times \mathbb{R}$. In fact, each reference frame F attaches a Cartesian system of coordinates to the space-time points, and any two inertial reference frames F, F' are obtained from each other by affecting a rotation R, a velocity boost $\boldsymbol{v}$, a space translation $\boldsymbol{a}$, and a time translation b. In other words, the coordinates $(\boldsymbol{x}', t')$ of an inertial frame F' are related to the coordinates $(\boldsymbol{x}, t)$ of another inertial frame F as follows:

$$\begin{aligned} \boldsymbol{x}' &= R\boldsymbol{x} + \boldsymbol{v}t + \boldsymbol{a}, \\ t' &= b + t. \end{aligned}$$

This allows one to regard the Galilei group G as the group of the coordinate transformations between the inertial frames of the Newtonian space-time.

4.1.2 The Covering Group

Regarding G as an analytic manifold, G is the product of $\mathcal{T} \times \mathcal{V} = \mathbb{R}^7$ and of $SO(3)$. The manifold $\mathbb{R}^7$ is simply connected, whereas $SO(3)$ fails to be simply connected and its universal covering group is the (complex) special unitary group $SU(2)$. We let $\delta : SU(2) \to SO(3)$ denote the covering homomorphism. It is an analytic function and its kernel is $\ker(\delta) = \{\pm I\}$. The group

$$G^* := \mathcal{T} \times' (\mathcal{V} \times' SU(2)) \equiv \mathcal{T} \times' G_o^*$$

is connected and simply connected and is the *covering group* of the Galilei group. The covering homomorphism $\delta : G^* \to G$ is given by $\delta((\boldsymbol{a}, b, \boldsymbol{v}, h)) = (\boldsymbol{a}, b, \boldsymbol{v}, \delta(h))$ and its kernel consists of the two elements $\pm e^*$.

4.1.3 The Lie Algebra

Since $\mathcal{T}$ and $\mathcal{V}$ are vector groups, their Lie algebras can be canonically identified with the vector spaces $\mathbb{R}^4$ and $\mathbb{R}^3$, respectively,

$$\begin{aligned}\mathrm{Lie}\,(\mathcal{T}) &= \mathbb{R}^4,\\ \mathrm{Lie}\,(\mathcal{V}) &= \mathbb{R}^3.\end{aligned}$$

The exponential maps $\mathrm{Lie}\,(\mathcal{T}) \to \mathcal{T}$ and $\mathrm{Lie}\,(\mathcal{V}) \to \mathcal{V}$ are then the identity maps. These Lie algebras are Abelian, so that, for instance, $[a_1, a_2] = 0$ for all $a_1, a_2 \in \mathrm{Lie}\,(\mathcal{T})$. The Lie algebra of $SO(3)$ is the vector space $so\,(3)$ of 3×3 real traceless skew symmetric matrices,

$$\mathrm{Lie}\,(SO(3)) = so\,(3).$$

The exponential map $\exp : so\,(3) \to SO(3)$ is the usual exponential map of matrices, $\exp A = e^A$, $A \in so\,(3)$, and the bracket is now the commutator of the matrices, $[A_1, A_2] = A_1A_2 - A_2A_1$, $A_1, A_2 \in so\,(3)$. The Lie algebra $so\,(3)$ is not Abelian and it has no proper ideals, that is, $so\,(3)$ is simple. Therefore, the linear span of the elements $[A_1, A_2], A_1, A_2 \in so\,(3)$, is the whole $so\,(3)$,

$$[so\,(3), so\,(3)] = so\,(3).$$

The Lie algebra of $G_0 = \mathcal{V} \times' SO(3)$, as a vector space, is

$$\begin{aligned}\mathrm{Lie}\,(G_0) &= \mathrm{Lie}\,(\mathcal{V}) \oplus \mathrm{Lie}\,(SO(3))\\ &= \mathbb{R}^3 \oplus so\,(3).\end{aligned}$$

Due to the action of $SO(3)$ on $\mathcal{V}$, $R[\boldsymbol{v}] = R\boldsymbol{v}$, one has $[\boldsymbol{v}, A] = A\boldsymbol{v}$ for all $\boldsymbol{v} \in \mathrm{Lie}\,(\mathcal{V}), A \in so\,(3)$, showing that $\mathrm{Lie}\,(\mathcal{V})$ is an ideal in $\mathrm{Lie}\,(G_0)$; in fact,

$$[\mathrm{Lie}\,(\mathcal{V}), so\,(3)] = \mathrm{Lie}\,(\mathcal{V}).$$

The Lie algebra of $G = \mathcal{T} \times' G_0$, as a vector space, is

$$\begin{aligned}\mathrm{Lie}\,(G) &= \mathrm{Lie}\,(\mathcal{T}) \oplus \mathrm{Lie}\,(\mathcal{V}) \oplus \mathrm{Lie}\,(SO(3))\\ &= \mathbb{R}^4 \oplus \mathbb{R}^3 \oplus so\,(3).\end{aligned}$$

We let $X = (a, \boldsymbol{v}, A) \equiv (\boldsymbol{a}, b, \boldsymbol{v}, A)$ denote its generic element. Taking into account the semidirect product structure of G one obtains

$$[\mathrm{Lie}\,(\mathcal{T}), \mathrm{Lie}\,(G_0)] \subset \mathrm{Lie}\,(\mathcal{T}).$$

Since $\mathcal{T}$ is normal in G, the inner automorphisms define a natural action of $SO(3)$ on $\mathcal{T}$, which preserves the splitting $\mathcal{T} = \mathcal{T}_s \times \mathcal{T}_t$. Taking the adjoint action on $\mathrm{Lie}\,(\mathcal{T}) = \mathrm{Lie}\,(\mathcal{T}_s) \oplus \mathrm{Lie}\,(\mathcal{T}_t)$, we have the brackets $[\boldsymbol{a}, A] = A\boldsymbol{a}$, $\boldsymbol{a} \in \mathrm{Lie}\,(\mathcal{T}_s), A \in so\,(3)$, $[b, A] = 0$, $b \in \mathrm{Lie}\,(\mathcal{T}_t), A \in so\,(3)$, showing that

$$[\mathrm{Lie}\,(\mathcal{T}_s), so\,(3)] = \mathrm{Lie}\,(\mathcal{T}_s)$$
$$[\mathrm{Lie}\,(\mathcal{T}_t), so\,(3)] = 0.$$

Moreover, by explicit calculation, one gets $[b, \boldsymbol{v}] = b\boldsymbol{v}$ and $[\boldsymbol{a}, \boldsymbol{v}] = 0$, that is,

$$[\mathrm{Lie}\,(\mathcal{T}_t), \mathrm{Lie}\,(\mathcal{V})] = \mathrm{Lie}\,(\mathcal{T}_s)$$
$$[\mathrm{Lie}\,(\mathcal{T}_s), \mathrm{Lie}\,(\mathcal{V})] = 0.$$

In particular, the above relations show that $\mathrm{Lie}\,(\mathcal{T})$ is an ideal in $\mathrm{Lie}\,(G)$.

The Lie algebra of the covering group $G^* = \mathcal{T} \times' (\mathcal{V} \times' SU(2))$ of the Galilei group is

$$\begin{aligned}\mathrm{Lie}\,(G^*) &= \mathrm{Lie}\,(\mathcal{T}) \oplus \mathrm{Lie}\,(\mathcal{V}) \oplus \mathrm{Lie}\,(SU(2)) \\ &= \mathbb{R}^4 \oplus \mathbb{R}^3 \oplus su\,(2),\end{aligned}$$

where $su\,(2)$ is the vector space of all 2×2 skew Hermitian complex matrices. The covering homomorphism $\delta : SU(2) \to SO(3)$ induces an isomorphism between the Lie algebras of $SU(2)$ and $SO(3)$. We denote it by $\delta^* : su\,(2) \to so\,(3)$. The Lie algebras $\mathrm{Lie}\,(G)$ and $\mathrm{Lie}\,(G^*)$ of the Galilei group G and its covering group G^* are thus isomorphic. We denote an element of $\mathrm{Lie}\,(G^*)$ by $X^* = (a, \boldsymbol{v}, B) \equiv (\boldsymbol{a}, b, \boldsymbol{v}, B)$, $B \in su\,(2)$, and we recall that, for instance, $[\boldsymbol{v}, B] = \delta^*(B)\boldsymbol{v}$ and $[\boldsymbol{a}, B] = \delta^*(B)\boldsymbol{a}$.

4.1.4 The Multipliers for the Covering Group

We next compute the multipliers of the covering G^* of the Galilei group, which is a connected simply connected Lie group. Hence, the problem is reduced to one of studying the closed forms on its Lie algebra $\mathrm{Lie}\,(G^*)$.

Let F be a closed $\mathbb{R}$-form on the Lie algebra $\mathrm{Lie}\,(G^*)$. We observe that $su\,(2)$ acts on $\mathcal{V}$, endowed with the usual inner product, as an irreducible orthogonal representation and on $\mathcal{T}$, endowed with the usual inner product, as an orthogonal representation which is the direct sum of two non equivalent irreducible representations acting on $\mathcal{T}_s$ and $\mathcal{T}_t$, respectively.

Using Theorem 7.40 of [35] one concludes that F restricted to $\mathrm{Lie}\,(G_0^*) \times \mathrm{Lie}\,(G_0^*)$ and $(\mathrm{Lie}\,(\mathcal{T}) \oplus su\,(2)) \times (\mathrm{Lie}\,(\mathcal{T}) \oplus su\,(2))$ is exact, so that there is a linear function q_1: $\mathrm{Lie}\,(G^*) \to \mathbb{R}$ such that

$$F(X, Y) = q_1([X, Y]),\;\; X, Y \in \mathrm{Lie}\,(G_0^*),$$
$$q_1(a) = 0,\;\; a \in \mathrm{Lie}\,(\mathcal{T}).$$

Let $F'(X, Y) := F(X, Y) - q_1([X, Y])$, $X, Y \in \mathrm{Lie}\,(G^*)$. Then F' is equivalent to F and it is zero on $\mathrm{Lie}\,(G_0^*) \times \mathrm{Lie}\,(G_0^*)$. Moreover, there is a linear function q_2 on $\mathrm{Lie}\,(G^*)$ such that

$$F'(X,Y) = q_2([X,Y]),\ X,Y \in \mathrm{Lie}\,(\mathcal{T}) \oplus su\,(2),$$
$$q_2(\boldsymbol{v}) = 0,\ \boldsymbol{v} \in \mathrm{Lie}\,(\mathcal{V}).$$

Let $F''(X,Y) := F'(X,Y) - q_2([X,Y]),\ X,Y \in \mathrm{Lie}\,(G^*)$. Then F'' is equivalent to F and it is zero on $(\mathrm{Lie}\,(\mathcal{T}) \oplus su\,(2)) \times (\mathrm{Lie}\,(\mathcal{T}) \oplus su\,(2))$ as well as on $\mathrm{Lie}\,(G_0^*) \times \mathrm{Lie}\,(G_0^*)$ since $[\mathrm{Lie}\,(\mathcal{V}), \mathrm{Lie}\,(G_0^*)] \subset \mathrm{Lie}\,(\mathcal{V})$.

Let $b \in \mathrm{Lie}\,(\mathcal{T}_t)$ and $\boldsymbol{v} \in \mathrm{Lie}\,(\mathcal{V})$. Since $\mathrm{Lie}\,(\mathcal{V}) = [\mathrm{Lie}\,(\mathcal{V}), su\,(2)]$ there exist $\boldsymbol{v}' \in \mathrm{Lie}\,(\mathcal{V})$ and $B \in su\,(2)$ such that $\boldsymbol{v} = \delta^*(B)\boldsymbol{v}' = [\boldsymbol{v}', B]$. Hence,

$$F''(b,\boldsymbol{v}) = F''(b,[\boldsymbol{v}',B]) = -F''(B,[b,\boldsymbol{v}']) - F''(\boldsymbol{v}',[B,b]) = 0$$

since $[b,\boldsymbol{v}'] = b\boldsymbol{v}' \in \mathrm{Lie}\,(\mathcal{T}_s)$ and $[B,b] = 0$.

We are left with the restriction of F'' to $\mathrm{Lie}\,(\mathcal{V}) \times \mathrm{Lie}\,(\mathcal{T}_s)$, that is, to $\mathbb{R}^3 \times \mathbb{R}^3$. Let C be the operator on $\mathbb{R}^3$ such that

$$F''(\boldsymbol{v},\boldsymbol{a}) = \boldsymbol{v} \cdot C\boldsymbol{a},\ \boldsymbol{v} \in \mathrm{Lie}\,(\mathcal{V}),\ \boldsymbol{a} \in \mathrm{Lie}\,(\mathcal{T}_s).$$

Since for all $B \in su\,(2)$

$$F''(\boldsymbol{v},[\boldsymbol{a},B]) + F''(\boldsymbol{a},[B,\boldsymbol{v}]) = 0,\ \boldsymbol{v} \in \mathrm{Lie}\,(\mathcal{V}),\ \boldsymbol{a} \in \mathrm{Lie}\,(\mathcal{T}_s),$$

one obtains that $C\delta^*(B) = \delta^*(B)C^t,\ B \in su\,(2)$, where C^t denotes the transpose of C. This then implies that C is a multiple of the identity operator.

Collecting all the results, we have that, given a closed form $F : \mathrm{Lie}\,(G^*) \times \mathrm{Lie}\,(G^*) \to \mathbb{R}$, there is a real number $m \in \mathbb{R}$ such that F is equivalent to F_m, where F_m is given by

$$F_m(X_1,X_2) = m(\boldsymbol{v}_1 \cdot \boldsymbol{a}_2 - \boldsymbol{a}_1 \cdot \boldsymbol{v}_2),$$

with $X_i = (\boldsymbol{a}_i, b_i, \boldsymbol{v}_i, B_i) \in \mathrm{Lie}\,(G^*), i = 1,2$. The form F_m is exact if and only if $m = 0$ and the forms F_m and $F_{m'}$ are equivalent if and only if $m = m'$.

It follows that the vector space $H^2(\mathrm{Lie}\,(G^*),\mathbb{R})$ is one dimensional and its elements are the equivalence classes $[F_m]$, $m \in \mathbb{R}$.

4.1.5 The Universal Central Extension

We close this section by computing the universal central extension of the covering group of the Galilei group. Since $H^2(\mathrm{Lie}\,(G^*),\mathbb{R})$ is one dimensional, $\overline{G}$ is of the form $\mathbb{R} \times'_\tau G^*$, where τ is an analytic $\mathbb{R}$-multiplier for G^* such that $[F_1] = [F_\tau]$, as given in Lemma 8 of Sect. 3.2 of Chap. 3, and

$$F_1(X_1^*,X_2^*) = \boldsymbol{v}_1 \cdot \boldsymbol{a}_2 - \boldsymbol{a}_1 \cdot \boldsymbol{v}_2,$$

with $X_i^* = (\boldsymbol{a}_i, b_i, \boldsymbol{v}_i, B_i) \in \mathrm{Lie}\,(G^*), i = 1,2$.

To compute τ, we proceed to show that $\overline{G}$ is, in fact, a semidirect product of $\mathbb{R} \times \mathcal{T} = \mathbb{R}^5$ and G_0^*. This can be done by studying the Lie algebra of $\overline{G}$.

The Lie algebra of $\overline{G}$ is given by

$$\begin{aligned}\mathrm{Lie}\,(\overline{G}) &= \mathbb{R} \oplus \mathrm{Lie}\,(G^*) \\ &= \mathbb{R} \oplus (\mathrm{Lie}\,(\mathcal{T}) \oplus (\mathrm{Lie}\,(\mathcal{V}) \oplus su\,(2))),\end{aligned}$$

with the Lie product

$$[(m_1, X_1^*), (m_2, X_2^*)] := (F_1(X_1^*, X_2^*), [X_1^*, X_2^*]),$$
$$m_1, m_2 \in \mathbb{R}, X_1^*, X_2^* \in \mathrm{Lie}\,(G^*).$$

By a direct computation one can confirm that $\mathrm{Lie}\,(\mathcal{V}) \oplus su\,(2)$ is a subalgebra of $\mathrm{Lie}\,(\overline{G})$,

$$[\mathrm{Lie}\,(\mathcal{V}) \oplus su\,(2), \mathrm{Lie}\,(\mathcal{V}) \oplus su\,(2)] \subseteq \mathrm{Lie}\,(\mathcal{V}) \oplus su\,(2),$$

with the identification of an element $(\boldsymbol{v}, B) \in \mathrm{Lie}\,(\mathcal{V}) \oplus su\,(2)$ with $(0, \boldsymbol{v}, B) \in \mathrm{Lie}\,(\overline{G}), 0 \in \mathbb{R}^5$. As another immediate observation one has that $\mathbb{R} \oplus \mathrm{Lie}\,(\mathcal{T}) = \mathbb{R}^5$ is an Abelian subalgebra of $\mathrm{Lie}\,(\overline{G})$, again, with the identification of $(m, \boldsymbol{a}, b) \in \mathbb{R} \oplus \mathrm{Lie}\,(\mathcal{T})$ with $(m, \boldsymbol{a}, b, \boldsymbol{0}, 0) \in \mathrm{Lie}\,(\overline{G})$. In fact, $\mathbb{R} \oplus \mathrm{Lie}\,(\mathcal{T})$ is an ideal of $\mathrm{Lie}\,(\overline{G})$. Indeed, for $(\boldsymbol{v}, B) \in \mathrm{Lie}\,(\mathcal{V}) \oplus su\,(2)$ and $(m, \boldsymbol{a}, b) \in \mathbb{R} \oplus \mathrm{Lie}\,(\mathcal{T})$ one has

$$\begin{aligned}[(\boldsymbol{v}, B), (m, \boldsymbol{a}, b,)] &\equiv [(0, \boldsymbol{0}, 0, \boldsymbol{v}, B), (m, \boldsymbol{a}, b, \boldsymbol{0}, 0)] \\ &= (\boldsymbol{v} \cdot \boldsymbol{a}, [(\boldsymbol{v}, B), (\boldsymbol{a}, b)]) \\ &= (\boldsymbol{v} \cdot \boldsymbol{a}, \delta^*(B)\boldsymbol{a} + b\boldsymbol{v}, 0, \boldsymbol{0}, 0) \\ &=: \rho^*(\boldsymbol{v}, B)(m, \boldsymbol{a}, b),\end{aligned}$$

with $\rho^*(\boldsymbol{v}, B)$ denoting the 5×5 real matrix

$$\rho^*(\boldsymbol{v}, B) = \begin{pmatrix} 0 & \boldsymbol{v} & 0 \\ \boldsymbol{0} & \delta^*(B) & \boldsymbol{v} \\ 0 & 0 & 0 \end{pmatrix},$$

which acts on the (column) vector $(m, \boldsymbol{a}, b) \in \mathbb{R}^5$. This shows that $\mathrm{Lie}\,(\overline{G})$ is a semidirect product of $\mathbb{R} \oplus \mathrm{Lie}\,(\mathcal{T})$ with $\mathrm{Lie}\,(\mathcal{V}) \oplus su\,(2)$ relative to ρ^*. Therefore, $\overline{G}$ is a semidirect product of $\mathbb{R} \times \mathcal{T}$ and $\mathcal{V} \times' SU(2)$, and it remains to determine the action of $\mathcal{V} \times' SU(2)$ on $\mathbb{R} \times \mathcal{T}$.

The action of $(\boldsymbol{v}, h) \in \mathcal{V} \times' SU(2)$ on $\mathbb{R}^5$ is given by the 5×5 matrix $\rho(\boldsymbol{v}, h)$ such that the differential at the identity of $\rho : (\boldsymbol{v}, h) \mapsto \rho(\boldsymbol{v}, h)$ is $\rho^*(\boldsymbol{v}, B)$. Since ρ is a representation we can compute the action of $\boldsymbol{v}$ and h separately.

Let $B \in \mathrm{Lie}\,(su\,(2))$. Then

$$\begin{aligned}\rho(\boldsymbol{0}, e^B) &= e^{\rho^*(\boldsymbol{0}, B)} \\ &= \sum_{n=0}^{\infty} \frac{1}{n!} \rho^*(\boldsymbol{0}, B) \\ &= \begin{pmatrix} 1 & \boldsymbol{0} & 0 \\ \boldsymbol{0} & e^{\delta^*(B)} & \boldsymbol{0} \\ 0 & \boldsymbol{0} & 1 \end{pmatrix}.\end{aligned}$$

Since $e^{\delta^*(B)} = \delta(e^B)$ and the exponential map is surjective we have for $h = e^B$

$$\rho(\mathbf{0}, h) = \begin{pmatrix} 1 & \mathbf{0} & 0 \\ \mathbf{0} & \delta(h) & \mathbf{0} \\ 0 & \mathbf{0} & 1 \end{pmatrix} .$$

Let $\boldsymbol{v} \in \mathrm{Lie}\,(\mathcal{V})$ so that $\exp \boldsymbol{v} = \boldsymbol{v} \in \mathcal{V}$. By a direct computation one now gets

$$\rho(\boldsymbol{v}, I) = e^{\rho^*(\boldsymbol{v},0)} = \begin{pmatrix} 1 & \boldsymbol{v} & \frac{1}{2}\boldsymbol{v}^2 \\ \mathbf{0} & I & \boldsymbol{v} \\ 0 & 0 & 1 \end{pmatrix} .$$

Therefore,

$$\rho(\boldsymbol{v}, h) = \rho(\boldsymbol{v}, I)\rho(\mathbf{0}, h) = \begin{pmatrix} 1 & \boldsymbol{v}\cdot\delta(h) & \frac{1}{2}\boldsymbol{v}^2 \\ \mathbf{0} & \delta(h) & \boldsymbol{v} \\ 0 & 0 & 1 \end{pmatrix} .$$

The action of $(\boldsymbol{v}, h) \in \mathcal{V} \times' SU(2)$ on $\mathbb{R}^5$ is thus explicitly given by

$$\rho(\boldsymbol{v}, h)(m, \boldsymbol{a}, b) = (m + \boldsymbol{v}\cdot\delta(h)\boldsymbol{a} + \tfrac{1}{2}b\boldsymbol{v}^2, \delta(h)\boldsymbol{a} + b\boldsymbol{v}, b).$$

From that one may also extract the corresponding multiplier for G^*. Indeed, for any $(0, \boldsymbol{a}_i, b_i, \boldsymbol{v}_i, h_i) \in \overline{G}$, $i = 1, 2$, the multiplication law is now given as

$$\begin{aligned} &(0, \boldsymbol{a}_1, b_1, \boldsymbol{v}_1, h_1)(0, \boldsymbol{a}_2, b_2, \boldsymbol{v}_2, h_2) = \\ &\quad (\boldsymbol{v}_1\cdot\delta(h_1)\boldsymbol{a}_2 + \tfrac{1}{2}b_2\boldsymbol{v}^2, \boldsymbol{a}_1 + \delta(h_1)\boldsymbol{a}_2 + b_2\boldsymbol{v}_1, b_1 + b_2, \boldsymbol{v} + \delta(h_1)\boldsymbol{v}_2, h_1h_2), \end{aligned}$$

which shows that

$$\tau(g_1^*, g_2^*) = \boldsymbol{v}_1\cdot\delta(h_1)\boldsymbol{a}_2 + \tfrac{1}{2}b_2\boldsymbol{v}_1^2.$$

Since $\overline{G} = \mathbb{R} \times'_{\tau} G^*$, the group law in $\overline{G}$ is now given as follows: for any $(c_1, g_1^*), (c_2, g_2^*) \in \overline{G}$, with $g_i^* = (\boldsymbol{a}_i, b_i, \boldsymbol{v}_i, h_i) \in G^*, c_i \in \mathbb{R}$,

$$\begin{aligned} (c_1, g_1^*)(c_2, g_2^*) &= (c_1 + c_2 + \tau(g_1^*, g_2^*), g_1^*g_2^*) \\ &= (c_1 + c_2 + \tfrac{1}{2}b_2\boldsymbol{v}_1^2 + \delta(h_1)\boldsymbol{a}_2\cdot\boldsymbol{v}_1, g_1^*g_2^*), \end{aligned}$$

where we have fixed the vector $[F_1]$ as the basis of the vector space H^2 $(\mathrm{Lie}\,(G^*), \mathbb{R})$.

Recalling that $G^* = \mathcal{T} \times' (\mathcal{V} \times' SU(2))$, we observe first that $A \equiv \mathbb{R}^5 = \{(\boldsymbol{a}, b, c) : (\boldsymbol{a}, b) \in \mathcal{T}, c \in \mathbb{R}\}$, with the identification $(\boldsymbol{a}, b, c) = (c, \boldsymbol{a}, b, \mathbf{0}, I)$, and $H \equiv \mathcal{V}\times' SU(2)$, with the identification $(\boldsymbol{v}, h) = (\mathbf{0}, 0, \boldsymbol{v}, h, 0)$, are (closed Lie) subgroups of $\overline{G}$, A being a normal Abelian subgroup of it. By a direct computation one verifies that $\overline{G} = AH$ and $A \cap H$ contains only the identity element $(\mathbf{0}, 0, \mathbf{0}, I, 0)$ of $\overline{G}$. In other words,

$$\overline{G} = A \times' H,$$

and the action of H on A is given as

$$\begin{aligned} (\boldsymbol{v}, h)[a] &= (\boldsymbol{v}, h)(\boldsymbol{a}, b, c)(\boldsymbol{v}, h)^{-1} \\ &= (\delta(h)\boldsymbol{a} + b\boldsymbol{v}, b, c + \tfrac{1}{2}b\boldsymbol{v}^2 + \delta(h)\boldsymbol{a}\cdot\boldsymbol{v}). \end{aligned}$$

4.2 The 2 + 1 Dimensional Case

From the physical point of view, the interest in the Galilei group in $2+1$ dimensions arises in solid state physics where some genuine examples of two dimensional systems can be found.

In this section we use the vector notation $\boldsymbol{x}$ for the elements of $\mathbb{R}^2$. The Galilei group in $2+1$ dimensions is

$$G = \mathcal{T} \times' (\mathcal{V} \times' SO(2)),$$

where $\mathcal{T} = \mathcal{T}_s \times \mathcal{T}_t$, $\mathcal{T}_s = \mathbb{R}^2$, $\mathcal{T}_t = \mathbb{R}$, and $\mathcal{V} = \mathbb{R}^2$. The semidirect product structure is analogous to the $3+1$ dimensional case. The covering group is $G^* = \mathcal{T} \times' (\mathcal{V} \times' \mathbb{R})$ and we denote its element as $(\boldsymbol{a}, b, \boldsymbol{v}, r)$, where $\boldsymbol{a}, \boldsymbol{v} \in \mathbb{R}^2$, $b \in \mathbb{R}$ and $r \in \mathbb{R}$. The kernel of the covering homomorphism δ is

$$\{(\boldsymbol{0}, 0, \boldsymbol{0}, 2\pi k) \ : \ k \in \mathbb{Z}\}.$$

The Lie algebra of G^* is, as a vector space,

$$\begin{aligned} \mathrm{Lie}\,(G^*) &= \mathrm{Lie}\,(\mathcal{T}) \oplus \mathrm{Lie}\,(\mathcal{V}) \oplus \mathrm{Lie}\,(\mathbb{R}) \\ &= \mathbb{R}^3 \oplus \mathbb{R}^2 \oplus \mathbb{R}\,. \end{aligned}$$

We denote the elements of $\mathrm{Lie}\,(G^*)$ by $(\boldsymbol{a}, b, \boldsymbol{v}, r)$, with $b, r \in \mathbb{R}$, $\boldsymbol{a}, \boldsymbol{v} \in \mathbb{R}^2$.

4.2.1 The Multipliers for the Covering Group and the Universal Central Extension

A result of Bose [4] shows that $H^2(\mathrm{Lie}\,(G^*), \mathbb{R})$ is a three dimensional vector space and a basis is given by the equivalence classes of the following closed $\mathbb{R}$-forms:

$$\begin{aligned} F_1\left((\boldsymbol{a}_1, b_1, \boldsymbol{v}_1, r_1), (\boldsymbol{a}_2, b_2, \boldsymbol{v}_2, r_2)\right) &= r_1 b_2 - r_2 b_1, \\ F_2\left((\boldsymbol{a}_1, b_1, \boldsymbol{v}_1, r_1), (\boldsymbol{a}_2, b_2, \boldsymbol{v}_2, r_2)\right) &= \boldsymbol{v}_1 \cdot \boldsymbol{a}_2 - \boldsymbol{v}_2 \cdot \boldsymbol{a}_1, \\ F_3\left((\boldsymbol{a}_1, b_1, \boldsymbol{v}_1, r_1), (\boldsymbol{a}_2, b_2, \boldsymbol{v}_2, r_2)\right) &= \boldsymbol{v}_1 \wedge \boldsymbol{v}_2, \end{aligned}$$

where $\boldsymbol{v}_1 \wedge \boldsymbol{v}_2$ is a shorthand notation for $v_{1x}v_{2y} - v_{2x}v_{1y}$. Define F as the closed $\mathbb{R}^3$-form $F = (F_1, F_2, F_3)$. To compute the corresponding $\mathbb{R}^3$-multiplier τ_F of G^*, we have to determine the simply connected Lie group G^*_F with Lie algebra

$$\mathrm{Lie}\,(G^*_F) = \mathbb{R}^3 \oplus_F \mathrm{Lie}\,(G^*).$$

The algebra $\mathrm{Lie}\,(G^*_F)$ is, in fact, a semidirect sum. This can be seen as follows. Write $\mathrm{Lie}\,(G^*_F) = \mathbb{R}^2 \oplus \mathrm{Lie}\,(G^*) \oplus \mathbb{R}$ and its elements as (c_1, c_2, X, x) with $c_1, c_2, x \in \mathbb{R}$ and $X \in \mathrm{Lie}\,(G^*)$ such that

$$[(c_1, c_2, X, x), (c_1, c_2, X', x')] = (\,F_1(X, X'), F_2(X, X'), [X, X'], F_3(X, X')\,)\,.$$

By direct computation, the set

$$\{(\boldsymbol{v}, r, x) \equiv (0, 0, \mathbf{0}, 0, \boldsymbol{v}, r, x) \; : \; (\boldsymbol{v}, r, x) \in \mathrm{Lie}\,(\mathcal{V}) \oplus \mathrm{Lie}\,(\mathbb{R}) \oplus \mathbb{R}\}$$

is a subalgebra of $\mathrm{Lie}\,(G_F^*)$ with Lie brackets

$$[(\boldsymbol{v}, r, x), (\boldsymbol{v}', r', x')] = (\dot{\delta}(r)\boldsymbol{v}, 0, \boldsymbol{v} \wedge \boldsymbol{v}') \,,$$

where $(\boldsymbol{v}, r, x), (\boldsymbol{v}', r', x') \in \mathrm{Lie}\,(\mathcal{V}) \oplus \mathrm{Lie}\,(\mathbb{R}) \oplus \mathbb{R}$. If $H = \mathcal{V} \times \mathbb{R} \times \mathbb{R}$ is the Lie group with the product

$$(\boldsymbol{v}, r, x)(\boldsymbol{v}', r', x') = (\boldsymbol{v} + \delta(r)\boldsymbol{v}', r + r', x + x' + \boldsymbol{v} \wedge \delta(r)\boldsymbol{v}'),$$

one can check that its Lie algebra is $\mathrm{Lie}\,(\mathcal{V}) \oplus \mathrm{Lie}\,(\mathbb{R}) \oplus \mathbb{R}$.

Moreover, the set

$$\{(c_1, c_2, \boldsymbol{a}, b) \equiv (c_1, c_2, \boldsymbol{a}, b, \mathbf{0}, 0, 0) \; : \; (c_1, c_2, \boldsymbol{a}, b) \in \mathbb{R}^2 \oplus \mathrm{Lie}\,(\mathcal{T})\}$$

is an Abelian ideal of $\mathrm{Lie}\,(G_F^*)$ isomorphic to $\mathrm{Lie}\,(\mathbb{R}^2 \times \mathcal{T})$.

Taking into account the previous results and the fact that, as a vector space,

$$\mathrm{Lie}\,(G_F^*) = \big(\mathbb{R}^2 \oplus \mathrm{Lie}\,(\mathcal{T})\big) \oplus (\mathrm{Lie}\,(\mathcal{V}) \oplus \mathrm{Lie}\,(\mathbb{R}) \oplus \mathbb{R}) \,,$$

the Lie algebra $\mathrm{Lie}\,(G_F^*)$ is isomorphic to the semidirect sum of $\mathrm{Lie}\,(\mathbb{R}^2 \times \mathcal{T})$ and $\mathrm{Lie}\,(H)$.

Explicitly, if $(\boldsymbol{v}, r, x) \in \mathrm{Lie}\,(H)$ and $(c_1, c_2, \boldsymbol{a}, b) \in \mathrm{Lie}\,(\mathbb{R}^2 \times \mathcal{T})$ one has

$$\begin{aligned}
[(\boldsymbol{v}, r, x), (c_1, c_2, \boldsymbol{a}, b)] &= (rb, \boldsymbol{v} \cdot \boldsymbol{a}, \dot{\delta}(r)\boldsymbol{a} + b\boldsymbol{v}, 0) \\
&=: \dot{\rho}(\boldsymbol{v}, r, x)(c_1, c_2, \boldsymbol{a}, b),
\end{aligned}$$

where $\dot{\rho}(\boldsymbol{v}, r, x)$ is the 5×5 matrix

$$\dot{\rho}(\boldsymbol{v}, r, x) = \begin{pmatrix} 0\,0 & \mathbf{0} & r \\ 0\,0 & \boldsymbol{v} & 0 \\ 0\,0 & \begin{matrix} 0 & -r \\ r & 0 \end{matrix} & \boldsymbol{v} \\ 0\,0 & \mathbf{0} & 0 \end{pmatrix},$$

which acts on the column vector $(c_1, c_2, \boldsymbol{a}, b) \in \mathrm{Lie}\,(\mathbb{R}^2 \times \mathcal{T})$.

If ρ is the representation of H such that its differential at the identity is $\dot{\rho}$, then G_F^* is the semidirect product of $\mathbb{R}^2 \times \mathcal{T}$ and H with respect to ρ. A simple calculation shows that

$$\rho(\boldsymbol{v},0,0)=\begin{pmatrix} 1 & 0 & \mathbf{0} & 0 \\ 0 & 1 & \boldsymbol{v} & \frac{1}{2}\boldsymbol{v}^2 \\ \begin{matrix}0\\0\end{matrix} & \begin{matrix}0\\0\end{matrix} & \begin{matrix}1 & 0\\ 0 & 1\end{matrix} & \boldsymbol{v} \\ 0 & 0 & \mathbf{0} & 1 \end{pmatrix}$$

$$\rho(\mathbf{0},r,0)=\begin{pmatrix} 1 & 0 & \mathbf{0} & r \\ 0 & 1 & \mathbf{0} & 0 \\ \begin{matrix}0\\0\end{matrix} & \begin{matrix}0\\0\end{matrix} & \delta(r) & \mathbf{0} \\ 0 & 0 & \mathbf{0} & 1 \end{pmatrix}$$

$$\rho(\mathbf{0},0,x)=\begin{pmatrix} 1 & 0 & \mathbf{0} & 0 \\ 0 & 1 & \mathbf{0} & 0 \\ \begin{matrix}0\\0\end{matrix} & \begin{matrix}0\\0\end{matrix} & \begin{matrix}1 & 0\\ 0 & 1\end{matrix} & \mathbf{0} \\ 0 & 0 & \mathbf{0} & 1 \end{pmatrix}.$$

Hence the action of H on $\mathbb{R}^2 \times \mathcal{T}$ is given by

$$(\boldsymbol{v},r,x)[(c_1,c_2,\boldsymbol{a},b)] = (c_1+br, c_2+\boldsymbol{v}\cdot\delta(r)\boldsymbol{a}+\frac{b\boldsymbol{v}^2}{2}, \delta(r)\boldsymbol{a}+b\boldsymbol{v}, b).$$

If $g=(c_1,c_2,\boldsymbol{a},b,\boldsymbol{v},r,x)$ and $g'=(c_1',c_2',\boldsymbol{a}',b',\boldsymbol{v}',r',x')$ are in G_F^*, then

$$\begin{aligned} gg' = (c_1+c_1'+b'r, c_2+c_2'+\boldsymbol{v}\cdot\delta(r)\boldsymbol{a}'+\frac{b'\boldsymbol{v}^2}{2}, \boldsymbol{a}+\delta(r)\boldsymbol{a}'+b'\boldsymbol{v}, b+b', \\ \boldsymbol{v}+\delta(r)\boldsymbol{v}', r+r', x+x'+\boldsymbol{v}\wedge\delta(r)\boldsymbol{v}'), \end{aligned}$$

so that the explicit form of $\tau_F=(\tau_1,\tau_2,\tau_3)$ is

$$\begin{aligned} \tau_1(g,g') &= b'r \\ \tau_2(g,g') &= \boldsymbol{v}\cdot\delta(r)\boldsymbol{a}'+b'\boldsymbol{v}^2/2 \\ \tau_3(g,g') &= \boldsymbol{v}\wedge\delta(r)\boldsymbol{v}'. \end{aligned}$$

By Theorem 2, the equivalence classes $[\tau_1]$, $[\tau_2]$, $[\tau_3]$ form a basis of $H^2(G^*,\mathbb{R})$. Moreover τ_2 and τ_3 satisfy the condition

$$\tau_i(k,g^*)=\tau_i(g^*,k), \qquad k\in \operatorname{Ker}\delta,\ g^*\in G^*,$$

while τ_1 does not. It follows that $\dim H^2(G^*,\mathbb{R})_\delta = 2$, $\overline{\tau}=(\tau_2,\tau_3)$ and the universal central extension $\overline{G}$ is of the form $\mathbb{R}^2\times_{\overline{\tau}}^{\prime} G^*$. We observe that $\overline{G}$ is the semidirect product of the closed vector subgroup

$$A=\mathcal{T}\times\mathbb{R}=\{(\boldsymbol{a},b,c)\simeq(c,\boldsymbol{a},b,\mathbf{0},0,0)\ :\ c\in\mathbb{R},\boldsymbol{a}\in\mathcal{T}_s,b\in\mathcal{T}_t\}$$

and the Lie subgroup H

$$H=\{(\boldsymbol{v},r,x)\simeq(0,\mathbf{0},0,\boldsymbol{v},r,x)\ :\ \boldsymbol{v}\in\mathcal{V},x,r\in\mathbb{R}\}$$

with respect to the action of H on A given by

$$(\boldsymbol{v}, r, x)[(\boldsymbol{a}, b, c)] = (\delta(r)\boldsymbol{a} + b\boldsymbol{v}, b, c + \boldsymbol{v} \cdot \delta(r)\boldsymbol{a} + \frac{b\boldsymbol{v}^2}{2}).$$

Finally, one has that

$$K = \{(\mathbf{0}, 0, c, \mathbf{0}, 2\pi n, x) \ : \ c, x \in \mathbb{R}, \ n \in \mathbb{Z}\} \simeq \mathbb{Z} \times \mathbb{R}^2.$$

5 Galilei Invariant Elementary Particles

In this chapter we apply the general theory of symmetry actions as developed in Chaps. 2 and 3 to the Galilei groups of Chap. 4 splitting the treatment again into the 3 + 1 and 2 + 1 dimensional cases. We find it worthwhile, however, to start with a general analysis of the constraints imposed by the relativity principle on the description of a physical system in quantum mechanics.

5.1 The Relativity Principle for Isolated Systems

The description of physical phenomena is always done according to the choice of a reference frame so that, in particular, the quantum mechanical description of a system is given with respect to a chosen frame. In the following we also use the term observer as a synonym for a reference frame. The relativity principle deals with the comparison of the descriptions of a physical system with respect to different observers. In order to do this one selects a preferred family of reference frames, namely the inertial observers.

The transformations between the space-time coordinates of inertial frames form a group G that we call the covariance group of space and time. Clearly G acts transitively and freely on the set of inertial observers. It is a basic experimental fact that this group is the Poincaré group though in many physical situations this group can also be approximated by the Galilei group. In view of this fact, we first discuss the relativity principle without specifying the explicit form of G.

Let $\mathcal{S}$ be a physical system. According to the general rules of quantum mechanics any inertial observer F describes $\mathcal{S}$ by fixing a Hilbert space $\mathcal{H}_F$ and identifying states and observables with suitable sets of operators on $\mathcal{H}_F$. It is no loss of generality to assume that the Hilbert space $\mathcal{H}_F$ does not depend on F, hence we denote by $\mathcal{H}$ a fixed complex separable Hilbert space used by all inertial observers to describe $\mathcal{S}$. We also assume that all inertial observers identify states and observables of $\mathcal{S}$ with operators on $\mathcal{H}$ according to their own coordinate system.

Moreover, we assume that the dynamical evolution of each observer preserves the natural structures of the sets of states and observables. This amounts to saying that, given any observer F, for all $t_1, t_2 \in \mathbb{R}$ there ex-

G. Cassinelli, E. De Vito, P.J. Lahti, and A. Levrero, *The Theory of Symmetry Actions in Quantum Mechanics*, Lect. Notes Phys. **654**, pp. 61–72
http://www.springerlink.com/

ist $D_F(t_1, t_2)$ satisfying

$$\begin{aligned} D_F(t_1, t_2) &\in \Sigma \\ D_F(t_2, t_3) D_F(t_1, t_2) &= D_F(t_1, t_3) \\ D_F(t_1, t_2) &= D_F(t_2, t_1)^{-1} \end{aligned}$$

for all $t_1, t_2, t_3 \in \mathbb{R}$. The mapping $D_F(t_1, t_2)$ represents time evolution from t_1 to t_2 according to F.

Any observer F associates to the evolving system $\mathcal{S}$ a *dynamical state*, namely a map

$$\mathbb{R} \ni t \mapsto T_F(t) \in \mathbf{S}$$

such that $T_F(t_2) = D_F(t_1, t_2) T_F(t_1)$ for all t_1, t_2 in $\mathbb{R}$. This map is completely determined by D_F and $T_F(0)$, that is the state that F assigns to $\mathcal{S}$ at the origin of his time. Hence, any observer identifies his set of dynamical states with the set of states $\mathbf{S}$ mapping each dynamical state to the corresponding initial state. Note that this identification depends on the observer F and on the dynamics.

Two inertial observers F and F' describe the same evolving system in general with two different initial states T and T' and this defines a bijective function $s_F^{F'} : \mathbf{S} \to \mathbf{S}$ mapping T to T'. It is natural to assume that $s_F^{F'}$ preserves the structure of the set of states, so that $s_F^{F'} \in \Sigma$. Moreover, by definition, $s_{F'}^{F''} s_F^{F'} = s_F^{F''}$ must hold for all inertial observers F, F', F''. Since the group G acts transitively and freely on the set of inertial observers, we will denote the map $s_F^{F'}$ as $s_{F,g}$, where $g \in G$ is the only element in G such that $gF = F'$. Hence we have

$$s_{gF,h} s_{F,g} = s_{F,hg} \,, \qquad g, h \in G. \tag{5.1}$$

It is important to note that this relation determines uniquely $s_{F,g}$ for all F and all g if $s_{F_0,g}$ is known for all g and for a fixed inertial observer F_0. In fact, suppose that $s_{F_0,g}$ is given for all g and let $F = hF_0$. If one defines $s_{F,g} = s_{F_0,gh} s_{F_0,h}^{-1}$, then, given $h' \in G$, one has

$$\begin{aligned} s_{h'F,g} &= s_{F_0,gh'h} s_{F_0,h'h}^{-1} \\ &= s_{F_0,gh'h} s_{F_0,h}^{-1} s_{F_0,h} s_{F_0,h'h}^{-1} \\ &= s_{hF_0,gh'} s_{hF_0,h'}^{-1} \\ &= s_{F,gh'} s_{F,h'}^{-1} \,, \end{aligned}$$

so that the maps $s_{F,g}$ satisfy (5.1).

The relativity principle states that all inertial observers are equivalent for the description of isolated systems. This principle is implemented in quantum mechanics in the following way: if $\mathcal{S}$ is an *isolated* system, the map $s_{F,g}$ depends only on the element g of the group G and not on the observer F.

From (5.1) and the relativity principle it follows that for an isolated system the map $G \ni g \mapsto s_g := s_{F,g} \in \Sigma$ is a group homomorphism (which does not depend on the choice of F). Moreover, it is natural to assume that this homomorphism is continuous, so that *the relativity principle associates to any isolated system a symmetry action of the space-time covariance group* G.

As a consequence of the relativity principle and the results obtained in Chap. 3, one sees that the *isolated and elementary* systems are described in quantum mechanics by the irreducible unitary representations of the universal central extension $\overline{G}$ of the covariance group G of space-time. In the following sections, we apply this fundamental result to the case of G being the Galilei group both in $3+1$ dimensions and in $2+1$ dimensions. However, to close this part, we discuss briefly the case of non isolated systems.

5.1.1 Galilei Systems in Interaction

For a *non isolated* system the maps $s_{F,g}$ depend in general both on F and on g. However in the case of Galilean relativity, the peculiar structure of the Galilei group allows us to guess the dependence of the maps $s_{F,g}$ from F and g also in the case of non isolated systems.

The Galilei group G can be written as

$$G = H \times' \mathcal{T}_t,$$

where $H = (\mathcal{T}_s \times \mathcal{V}) \times' SO(3)$ is the isochronous subgroup and $\mathcal{T}_t$ is the subgroup of time translations. The subgroup H is normal in G and $\mathcal{T}_t$ acts on H as

$$b[(\mathbf{a}, \mathbf{v}, R)] = (\mathbf{a} - b\mathbf{v}, \mathbf{v}, R).$$

Any element g of G can be written, in a unique way, as $g = hb$ with $h \in H$ and $b \in \mathcal{T}_t$.

Observe that, if b is an element of the time translation subgroup $\mathcal{T}_t$ and F is any inertial observer, by definition we have

$$s_{F,b} = D_F(0, -b), \tag{5.2}$$

where D_F is the time evolution operator of F.

Let now $\mathcal{S}$ be a physical system and assume it is isolated. As explained before, there is a symmetry action $g \mapsto s_g$ associated to $\mathcal{S}$. To stress the fact that s_g refers to a free system, we denote it by s_g^f. Suppose now that the same system is subject to an interaction. Then the maps $s_{F,g}$ will in general depend also on the observer F. However, fixing an inertial observer F_0 (the laboratory), if h is an element of the isochronous subgroup, F_0 and hF_0 have the same origin of time, so that it is natural to assume that the map $s_{F_0,h}$ is the same as in the free case, that is, we assume that

$$s_{F_0,h} = s_h^f, \qquad h \in H. \tag{5.3}$$

If $g = hb \in G$ then (5.1) implies $s_{F_0,g} = s_{F_0,hb} = s_{bF_0,h} s_{F_0,b}$. This suggests to assume that

$$s_{F_0,g} = s_h^f D_{F_0}(0,-b) \,, \qquad g = hb \in G. \tag{5.4}$$

As mentioned before, the above relations, together with (5.1) fix uniquely the maps $s_{F,g}$ for all F and g for the interacting system $\mathcal{S}$. Explicitly, let $g_1, g_2 \in G$, $g_1 = h_1 b_1$, $g_2 = h_2 b_2$ and consider the inertial observer $F = g_1 F_0$. Then

$$\begin{aligned} s_{F,g_2} &= s_{g_1 F_0, g_2} \\ &= s_{F_0, g_2 g_1} s_{F_0,g_1}^{-1} \\ &= s_{F_0, h_2 b_2 [h_1] b_2 b_1} s_{F_0,g_1}^{-1} \\ &= s^f_{h_2 b_2 [h_1]} D_{F_0}(0, -b_2 - b_1) D_{F_0}(0,-b_1)^{-1} s^f_{h_1^{-1}} \\ &= s^f_{h_2 b_2 [h_1]} D_{F_0}(-b_1, -b_2 - b_1) s^f_{h_1^{-1}} . \end{aligned}$$

This is consistent with (5.2) only if the dynamical evolution operators satisfy

$$D_F(0, b_2) = s^f_{b_2[h_1]} D_{F_0}(-b_1, -b_2 - b_1) s^f_{h_1^{-1}}$$

for $F = h_1 b_1 F_0$.

Moreover, one verifies that this construction is independent of the choice of the "laboratory frame" F_0. In particular, if $g_2 = h_2 \in H$,

$$s_{F,h_2} = s^f_{h_2 h_1} s^f_{h_1^{-1}} = s^f_{h_2} .$$

5.2 Symmetry Actions in 3 + 1 Dimensions

To determine the unitary irreducible representations of the universal central extension $\overline{G} = \mathbb{R}^5 \times' G_o^*$ of the Galilei group G we follow the prescription of Sects. 3.3 and 3.4 of Chap. 3.

5.2.1 The Dual Group and the Dual Action

Any quintuple $(\boldsymbol{p}, E, m)$ of real numbers defines a character χ of the vector group $A = \mathbb{R}^5$ through the formula:

$$\chi(\boldsymbol{a}, b, c) = e^{i(-\boldsymbol{p}\cdot\boldsymbol{a} + Eb + mc)} \,, \qquad (\boldsymbol{a}, b, c) \in \mathbb{R}^5$$

(for the sake of convenience, we have chosen a minus sign in the first term of the exponent). On the other hand, it is well known that all the characters of $\mathbb{R}^5$ are of this form. Therefore, the dual group $\hat{A}$ can be identified with the additive group $\mathbb{R}^5$, for which we use the notation $\mathbb{P}^5$. The elements of $\mathbb{P}^5$ are denoted by $p = (\boldsymbol{p}, E, m)$.

Since the action of an element $\bar{g} = (\boldsymbol{a}, b, c, \boldsymbol{v}, h) \in \overline{G}$ on $A = \mathbb{R}^5$ is given by

$$\bar{g}[(\boldsymbol{a}', b', c')] = (\delta(h)\boldsymbol{a}' + b'\boldsymbol{v}, b', c' + \tfrac{1}{2}b'\boldsymbol{v}^2 + \boldsymbol{v} \cdot \delta(h)\boldsymbol{a}'),$$

with $(\boldsymbol{a}', b', c') \in A$, the dual action on $\hat{A} = \mathbb{P}^5$ is

$$\bar{g}[(\boldsymbol{p}, E, m)] = (\delta(h)\boldsymbol{p} + m\boldsymbol{v}, E + \tfrac{1}{2}m\boldsymbol{v}^2 + \boldsymbol{v} \cdot \delta(h)\boldsymbol{p}, m),$$

with $(\boldsymbol{p}, E, m) \in \mathbb{P}^5$.

5.2.2 The Orbits and the Orbit Classes

The action of $\overline{G}$ on the dual group $\mathbb{P}^5$ splits it into three kinds of orbits.

For fixed $E_o, m \in \mathbb{R}$, $m \neq 0$, consider the point $p_{E_o,m} := (\boldsymbol{0}, E_o, m) \in \mathbb{P}^5$. Its orbit is

$$\overline{G}[p_{E_o,m}] = \{(\boldsymbol{p}, E, m) \,|\, \boldsymbol{p} \in \mathbb{P}^3, E = E_o + \frac{\boldsymbol{p}^2}{2m}\}.$$

Similarly, the orbit of a point $p_r := ((0, 0, r), 0, 0) \in \mathbb{P}^5$, with a fixed $r \in \mathbb{R}$, $r > 0$, is

$$\overline{G}[p_r] = \{(\boldsymbol{p}, E, 0) \,|\, \boldsymbol{p} \in \mathbb{P}^3, |\boldsymbol{p}| = r, E \in \mathbb{P}\}.$$

Finally, the orbit of a point $p_{E_o} := (\boldsymbol{0}, E_o, 0) \in \mathbb{P}^5$ is the singleton set

$$\overline{G}[p_{E_o}] = \{(\boldsymbol{0}, E_o, 0)\}.$$

By a direct inspection one observes that these three classes of orbits exhaust the whole set $\mathbb{P}^5$, and that these orbits are closed, and thus also locally closed in $\mathbb{P}^5$.

Any character of $\overline{G}$ is of the form $y\chi$, where $y = (\boldsymbol{0}, E_o, 0) \in \mathbb{P}^5$ and χ is a character of G_o^*. Since G_o^* is also a semidirect product, $G_o^* = \mathcal{V} \times' SU(2)$, its characters are also of the product form. However, $\hat{\mathcal{V}} = \mathbb{P}^3$ has only the origin as a one-point orbit and $SU(2)$, as a simple Lie group, has no nontrivial characters. Thus any character of $\overline{G}$ is of the form

$$\bar{g} = (\boldsymbol{a}, b, c, \boldsymbol{v}, h) \mapsto e^{ibE_o},$$

with $E_o \in \mathbb{P}$. The orbit classes are the following: for any $m > 0$,

$$\widetilde{\mathcal{O}}_m^1 = \cup_{E_o \in \mathbb{R}} \left(\overline{G}[p_{E_0,m}] \cup \overline{G}[p_{E_0,-m}] \right),$$

for any $r > 0$,

$$\widetilde{\mathcal{O}}_r^2 = \overline{G}[p_r],$$

and, finally,

$$\widetilde{\mathcal{O}^3} = \cup_{E_o \in \mathbb{R}} \overline{G}[(\boldsymbol{0}, E_o, 0)].$$

5.2.3 Representations Arising from $\widetilde{\mathcal{O}}^1_m$

In the orbit class $\widetilde{\mathcal{O}}^1_m$ we choose the orbit $\overline{G}[p_{0,m}]$. The function

$$\mathbb{P}^3 \ni \boldsymbol{p} \mapsto (\boldsymbol{p}, \tfrac{\boldsymbol{p}^2}{2m}, m) \in \overline{G}[p_{0,m}]$$

defines a global coordinate system (surjective diffeomorphism) of $\mathbb{P}^3$ onto $\overline{G}[p_{0,m}]$.

The dual action of $\overline{G}$ on $\overline{G}[p_{0,m}]$ induces an action on $\mathbb{P}^3$, for any $\bar{g} = (\boldsymbol{a}, b, c, \boldsymbol{v}, h) \in \overline{G}$ and for each $\boldsymbol{p} \in \mathbb{P}^3$,

$$\bar{g}[\boldsymbol{p}] = \delta(h)\boldsymbol{p} + m\boldsymbol{v}$$

(notice that the action of G^*_o is the natural action of the Euclidean group on $\mathbb{P}^3$). The Lebesgue measure $d\boldsymbol{p}$ is a $\overline{G}$-invariant σ-finite measure on $\mathbb{P}^3$ and the stability subgroup of the point $p_{0,m}$ of the orbit $G[p_{0,m}]$, that is, of the point $\boldsymbol{0} \in \mathbb{P}^3$, is readily seen to be $\overline{G}_{p_o} = A \times' SU(2)$. The map $\beta : \mathbb{P}^3 \to \overline{G}$

$$\boldsymbol{p} \mapsto (\boldsymbol{0}, 0, 0, \tfrac{\boldsymbol{p}}{m}, I),$$

has the properties $\beta(\boldsymbol{0}) = (\boldsymbol{0}, 0, 0, \boldsymbol{0}, I)$ and $\beta(\boldsymbol{p})[\boldsymbol{0}] = \boldsymbol{p}$, $\boldsymbol{p} \in \mathbb{P}^3$, showing that β is an (analytic) section for the action of $\overline{G}$ on $\overline{G}[p_{0,m}]$ and that β takes values in G^*_o. With this section one has that for each $\bar{g} = (\boldsymbol{a}, b, c, \boldsymbol{v}, h) \in \overline{G}$ and $\boldsymbol{p} \in \mathbb{P}^3$,

$$\beta(\boldsymbol{p})^{-1}\bar{g}\beta(\bar{g}^{-1}[\boldsymbol{p}]) = (\boldsymbol{a} - \frac{b\boldsymbol{p}}{m}, b, c + \frac{\boldsymbol{p}^2}{2m}b - \frac{\boldsymbol{p}\cdot\boldsymbol{a}}{m}, \boldsymbol{0}, h).$$

The group $SU(2)$ is a compact, connected, simply connected Lie group and it is well known that all its irreducible unitary representations are of the form $\mathbb{D}^j$ acting on the Hilbert space $\mathbb{C}^{2j+1}$, with $j = 0, \frac{1}{2}, 1, \frac{3}{2}, 2, \cdots$. Thus the $(p_m\mathbb{D}^j)$-induced irreducible unitary representations $U^{(m,j)}$ of $\overline{G}$ acting on $L^2(\mathbb{P}^3, d\boldsymbol{p}, \mathbb{C}^{2j+1})$ are of the form

$$(U^{(m,j)}_{\bar{g}} f)(\boldsymbol{p}) = e^{i(-\boldsymbol{p}\cdot\boldsymbol{a} + \frac{\boldsymbol{p}^2}{2m}b + mc)}\, \mathbb{D}^j(h)\, f(\delta(h^{-1})(\boldsymbol{p} - m\boldsymbol{v})),$$

for any $f \in L^2(\mathbb{P}^3, d\boldsymbol{p}, \mathbb{C}^{2j+1})$, $\boldsymbol{p} \in \mathbb{P}^3$, $\bar{g} = (\boldsymbol{a}, b, c, \boldsymbol{v}, h) \in \overline{G}$. According to the results of Sect. 3.3.4, the representations $U^{(m,j)}$, $m > 0, j = 0, \frac{1}{2}, 1, \frac{3}{2}, 2, \cdots$ are all physically inequivalent representations arising from the orbit classes $\widetilde{\mathcal{O}}^1_m$, $m > 0$.

5.2.4 Representations Arising from the Orbit Class $\widetilde{\mathcal{O}}^2_r$

Consider next the orbit $\overline{G}[p_r]$, and let S_1 denote the unit sphere centered at the origin of $\mathbb{P}^3$. The function

$$S_1 \times \mathbb{P} \ni (\boldsymbol{u}, E) \mapsto (r\boldsymbol{u}, E, 0) \in \overline{G}[p_r]$$

is a diffeomorphism allowing one to identify the orbit $\overline{G}[p_r]$ with the manifold $S_1 \times \mathbb{P}$. The action of $\overline{G}$ on $\overline{G}[p_r]$ can again be transferred to an action of $\overline{G}$ on $S_1 \times \mathbb{P}$,

$$(\boldsymbol{a}, b, c, \boldsymbol{v}, h)[(\boldsymbol{u}, E)] = (\delta(h)\boldsymbol{u}, E + r\boldsymbol{v} \cdot \delta(h)\boldsymbol{u}),$$

where $\bar{g} = (\boldsymbol{a}, b, c, \boldsymbol{v}, h) \in \overline{G}$ and $(\boldsymbol{u}, E) \in S_1 \times \mathbb{P}$.

Let $d\Omega$ denote the unique normalized rotation invariant measure on the sphere S_1 and let dE denote the Lebesgue measure on $\mathbb{P}$. The product measure $d\Omega dE$ is then a $\overline{G}$-invariant σ-finite measure on $S_1 \times \mathbb{P}$.

The point p_r of the orbit $\overline{G}[p_r]$ corresponds to, the point $((0,0,1),0) \in S_1 \times \mathbb{P}$. To determine the stabilizer of this point we observe that

$$(\boldsymbol{0}, 0, 0, \boldsymbol{v}, h)[((0,0,1),0)] = ((0,0,1),0)$$

if and only if $\delta(h)$ is a rotation of the sphere S_1 around its south-north axis and $\boldsymbol{v}$ is of the form $(v_1, v_2, 0)$. Since

$$\delta\begin{pmatrix} e^{it/2} & 0 \\ 0 & e^{-it/2} \end{pmatrix} = \begin{pmatrix} \cos t & \sin t & 0 \\ -\sin t & \cos t & 0 \\ 0 & 0 & 1 \end{pmatrix},$$

we observe that the stability subgroup of the point $((0,0,1),0) \in S_1 \times \mathbb{P}$ is

$$\begin{aligned} &A \times' E(2) = \\ &\quad A \times' \{(\boldsymbol{v}, h) \in G_o^* \,|\, \boldsymbol{v} = (v_1, v_2, 0), v_1, v_2 \in \mathbb{R}, h = \begin{pmatrix} z & 0 \\ 0 & z^{-1} \end{pmatrix}, z \in \mathbb{T}\}. \end{aligned}$$

Consider now the function $\beta : S_1 \times \mathbb{P} \to \overline{G}$,

$$(\boldsymbol{u}, E) \mapsto (\boldsymbol{0}, 0, 0, \tfrac{E}{r}\boldsymbol{u}, h_{\boldsymbol{u}}),$$

where $h_{\boldsymbol{u}} \in SU(2)$ is such that $\delta(h_{\boldsymbol{u}})(0,0,1) = \boldsymbol{u}$. Using the polar coordinates $\theta \in [0, \pi], \varphi \in [0, 2\pi)$ one has

$$\boldsymbol{u} = (\sin\theta\cos\varphi, \sin\theta\sin\varphi, \cos\theta)$$

so that

$$h_{\boldsymbol{u}} = \begin{pmatrix} e^{-i\varphi/2}\cos\frac{\theta}{2} & -e^{-i\varphi/2}\sin\frac{\theta}{2} \\ e^{i\varphi/2}\sin\frac{\theta}{2} & e^{i\varphi/2}\cos\frac{\theta}{2} \end{pmatrix}$$

and thus

$$\delta(h_{\boldsymbol{u}}) = \begin{pmatrix} \cos\theta\cos\varphi & -\sin\varphi & \sin\theta\cos\varphi \\ \cos\theta\sin\varphi & \cos\varphi & \sin\theta\sin\varphi \\ -\sin\theta & 0 & \cos\theta \end{pmatrix}.$$

Clearly, $\beta((0,0,1),0) = (\boldsymbol{0}, 0, 0, \boldsymbol{0}, I)$ and $\beta(\boldsymbol{u}, E)[((0,0,1),0)] = (\boldsymbol{u}, E)$ for all $(\boldsymbol{u}, E) \in S_1 \times \mathbb{P}$, showing that σ is an (analytic) section for the action of $\overline{G}$ on the orbit $\overline{G}[p_r]$ taking values in G_o^*. Using the polar coordinate representations of $\boldsymbol{u}, h_{\boldsymbol{u}}$, and $\delta(h_{\boldsymbol{u}})$, one may also easily compute that for any $\bar{g} = (\boldsymbol{a}, b, c, \boldsymbol{v}, h) \in \overline{G}$, $(\boldsymbol{u}, E) \in S_1 \times \mathbb{P}$,

$$\sigma(\boldsymbol{u}, E)^{-1}\, \bar{g}\, \sigma(\bar{g}^{-1}[(\boldsymbol{u}, E)]) =$$
$$(\sigma(\boldsymbol{u}, E)^{-1}[(\boldsymbol{a}, b, c)], -\boldsymbol{v} \cdot \boldsymbol{u}(0,0,1) + \delta(h_{\boldsymbol{u}}^{-1})\boldsymbol{v}, h_{\boldsymbol{u}}^{-1} h h_{\delta(h^{-1})\boldsymbol{u}}).$$

Since $(\delta(h_{\boldsymbol{u}}^{-1})\boldsymbol{v})_3 = \boldsymbol{u} \cdot \boldsymbol{v}$ and $\delta(h_{\boldsymbol{u}}^{-1} h h_{\delta(h^{-1})\boldsymbol{u}})(0,0,1) = (0,0,1)$ one confirms that

$$\sigma(\boldsymbol{u}, E)^{-1}\, \bar{g}\, \sigma(\bar{g}^{-1}[(\boldsymbol{u}, E)]) \in A \times' E(2)$$

for all $\bar{g} = (\boldsymbol{a}, b, c, \boldsymbol{v}, h) \in \overline{G}$.

Let L be a unitary irreducible representation of the stability subgroup $E(2)$ acting on a complex separable Hilbert space $\mathcal{K}$. The $(p_r L)$-induced representation $U^{(r,L)}$ of $\overline{G}$ then acts on $L^2(S_1 \times \mathbb{P}, d\Omega dE, \mathcal{K})$ according to

$$(U_{\bar{g}}^{(r,L)} f)(\boldsymbol{u}, E)$$
$$= e^{i(Eb - \boldsymbol{p}\cdot\boldsymbol{a})}\, L(\beta(\boldsymbol{u}, E)^{-1}\, (\boldsymbol{v}, h)\, \beta((\boldsymbol{v}, h)^{-1}[(\boldsymbol{u}, E)])) f(\delta(h^{-1})\boldsymbol{u}, E - \boldsymbol{v} \cdot \boldsymbol{u}),$$

for any $f \in L^2(S_1 \times \mathbb{P}, d\Omega dE, \mathcal{K})$, $(\boldsymbol{u}, E) \in S_1 \times \mathbb{P}$, $\bar{g} = (\boldsymbol{a}, b, c, \boldsymbol{v}, h) \in \overline{G}$.

To determine the representations L of $E(2)$ we exhibit first its semidirect product structure. For $\boldsymbol{v} = (v_1, v_2, 0) \in \mathbb{R}^3$ we define

$$\xi(\boldsymbol{v}) := v_1 - i v_2.$$

With this definition the product $(\boldsymbol{v}_1, h_1)(\boldsymbol{v}_2, h_2) = (\boldsymbol{v}, h)$ of two elements of $E(2)$ is given by

$$\xi(\boldsymbol{v}) = z_1^2 \xi(\boldsymbol{v}_2) + \xi(\boldsymbol{v}_1),\ h\ = \begin{pmatrix} z_1 z_2 & 0 \\ 0 & (z_1 z_2)^{-1} \end{pmatrix},$$

showing that $E(2)$ can be identified with the semidirect product $\mathbb{C} \times' \mathbb{T}$, with the multiplication $(\xi_1, z_1)(\xi_2, z_2) = (z_1^2 \xi_2 + \xi_1, z_1 z_2)$. The action of $\mathbb{T}$ on $\mathbb{C}$ is thus given by $z[\xi] = z^2 \xi$. The irreducible unitary representations of $E(2)$ can thus be induced from the irreducible unitary representations of the stability subgroups of the points of the orbits of $\hat{\mathbb{C}}$, the dual group of $\mathbb{C}$.

For any $w \in \mathbb{C}$, the mapping

$$x_w(\xi) := e^{i\,\mathrm{Re}\,(w\bar{\xi})}$$

is a character of the additive group of complex numbers $\mathbb{C}$, and all the characters of $\mathbb{C}$ are of this form. In fact, the map $\mathbb{C} \to \hat{\mathbb{C}}, w \mapsto x_w$ is a group isomorphism. The (dual) action of $\mathbb{T}$ on $\hat{\mathbb{C}}$ is easily seen to be $z[x_w] = x_{z^2 w}$ for all $z \in \mathbb{T}, w \in \mathbb{C}$. The orbits in $\hat{\mathbb{C}}$ are then the singleton set $\{0\}$ and the circles $\mathcal{O}_\rho := \{w \in \mathbb{C} \,|\, |w| = \rho\}$ $\rho > 0$. In the first case the stability subgroup is $\mathbb{T}$ itself. To find the stability subgroup for the case $\rho > 0$, fix the point

$w = \rho \in \mathcal{O}_\rho$. Since $z[\rho] = z^2\rho = \rho$ if and only if $z = \pm 1$, we observe that the stability subgroup of the point $w = \rho$ of the orbit $\mathcal{O}_\rho$ is the two-element group $Z^2 = \{1, -1\}$.

Consider the orbit $\{0\}$. The unitary irreducible representations of $\mathbb{T}$ are the characters $z \mapsto z^n$, $n \in \mathbb{Z}$. The induced unitary representations $L^{0,n}$ of $E(2)$ act on $\mathbb{C}$ as multiplications by z^n, that is, $L^{0,n}_{(\xi,z)} = M_{z^n}$.

Consider next an orbit $\mathcal{O}_\rho$, $\rho > 0$. The invariant measure on it is the normalized arc length $\frac{d\vartheta}{2\pi\rho}$, and the irreducible unitary representations of Z^2 are the trivial constant 1 representation and the one for which $1 \mapsto 1$, $-1 \mapsto -1$. The corresponding induced representations of $E(2)$ act on $L^2(\mathcal{O}_\rho, \frac{d\vartheta}{2\pi\rho}, \mathbb{C})$ and they are

$$(L^{+,\rho}_{(\xi,z)} f)(w) = e^{i\,\mathrm{Re}\,(w\overline{\xi})} f(z^{-2}w),$$
$$(L^{-,\rho}_{(\xi,z)} f)(w) = \pm e^{i\,\mathrm{Re}\,(w\overline{\xi})} f(z^{-2}w),$$

where the sign $\pm$ depends on the choice of the section c.

5.2.5 Representations Arising from the Orbit Class $\widetilde{\mathcal{O}}^3$

The stability subgroup of the orbit $\overline{G}[(\mathbf{0},0,0)] = \{(\mathbf{0},0,0)\}$ is the whole group G^*_o. Let Π be a unitary irreducible representation of G^*_o acting on $\mathcal{K}$. The Hilbert space of the induced representation $U^{0,\Pi}$ of $\overline{G}$ is then $L^2(\{(\mathbf{0},0,0)\}, \mu_{(\mathbf{0},0,0)}, \mathcal{K}) \simeq \mathcal{K}$ and it is defined by

$$U^{0,\Pi}_{(\boldsymbol{a},b,\boldsymbol{v},h,c)} f = \Pi(\boldsymbol{v},h)f, \quad f \in \mathcal{K}.$$

All the induced representations arising from the different orbits $G[(\mathbf{0},E,0)]$ in $\widetilde{\mathcal{O}}_{(\mathbf{0},0,0)}$ are physically equivalent.

5.3 Symmetry Actions in 2 + 1 Dimensions

5.3.1 Unitary Irreducible Representations of $\overline{G}$

Like in the 3+1 dimensional case the universal covering group $\overline{G}$ is a semidirect product, so that we may again apply the results of Sect. II. 3.4 to classify the irreducible inequivalent representations of $\overline{G}$. Let $\hat{A}$ be the dual group of A. We identify $\hat{A}$ with $\mathbb{P}^4$ using the pairing

$$\langle(\boldsymbol{p},E,m),(\boldsymbol{a},b,c)\rangle = -\boldsymbol{p}\cdot\boldsymbol{a} + Eb + mc.$$

The dual action of $\overline{G}$ on $\hat{A}$ is

$$g[(\boldsymbol{p},E,m)] = (\delta(r)\boldsymbol{p} + m\boldsymbol{v}, E + \delta(r)\boldsymbol{p}\cdot\boldsymbol{v} + \frac{1}{2}m\boldsymbol{v}^2, m),$$

where $g = (\boldsymbol{a},b,c,\boldsymbol{v},r,x) \in \overline{G}$. We have the following orbits for the dual action.

1. For each $l \in \mathbb{R}$, $l > 0$,
$$\overline{G}[(\boldsymbol{p}_l, 0, 0)] = \{(\boldsymbol{p}, E, 0) \ : \ \boldsymbol{p}^2 = l^2\},$$
where $\boldsymbol{p}_l = (0, l)$.
2. For each $E \in \mathbb{R}$,
$$\overline{G}[(\boldsymbol{0}, E, 0)] = \{(E, \boldsymbol{0}, 0)\}.$$
3. For each $m, E_o \in \mathbb{R}$, $m \neq 0$,
$$\overline{G}[(\boldsymbol{0}, E_o, m)] = \{(\boldsymbol{p}, E, m) \ : \ E - \frac{\boldsymbol{p}^2}{2m} = E_0\}.$$

All the orbits are closed in $\hat{A}$, hence the semidirect product is regular and Theorem 4 can be applied.

The set of singleton orbits is
$$\hat{A}_s = \{(\boldsymbol{0}, E, 0) \ : \ E \in \mathbb{R}\},$$
and the orbit classes of $\overline{G}$ are the following:

1. for each $l \in \mathbb{R}$, $l > 0$,
$$\widetilde{\mathcal{O}}_l^1 = \overline{G}[(\boldsymbol{p}_l, 0, 0)];$$
2.
$$\widetilde{\mathcal{O}}^2 = \bigcup_{E \in \mathbb{R}} \overline{G}[(\boldsymbol{0}, E, 0)];$$
3. for any $m > 0$,
$$\widetilde{\mathcal{O}}_m^3 = \bigcup_{E_o \in \mathbb{R}} \left(\overline{G}[(\boldsymbol{0}, E_o, m)] \cup \overline{G}[(\boldsymbol{p}, E_o, -m)]\right).$$

In the following we will exploit in detail only the third case, which presents some interesting physical features.

Let $m > 0$ and $p_m = (\boldsymbol{0}, 0, m) \in \widetilde{\mathcal{O}}_m^3$. The stability subgroup
$$\overline{G}_{p_m} \cap H = \{(\boldsymbol{v}, r, x) \in H \ : \ \boldsymbol{v} = \boldsymbol{0}\}$$
is isomorphic to $\mathbb{R}^2$ and its irreducible representations are its characters. Explicitly, $\lambda, \mu \in \mathbb{R}$ define the character of $\overline{G}_{p_m} \cap H$
$$(\boldsymbol{0}, r, x) \mapsto e^{i\lambda x} e^{i\mu r}.$$

Now we observe that

1. if $y \in \hat{A}_s$, $y \neq 0$, then $y\overline{G}[p_m] \neq \overline{G}[p_m]$;
2. $\overline{G}[p_m] \neq \overline{G}[p_m]^{-1}$;
3. the characters of H are of the form
$$(\boldsymbol{v}, r, x) \mapsto e^{i\mu r}.$$

According to Theorem 4, every irreducible representation of $\overline{G}$ living on an orbit class $\widetilde{\mathcal{O}}_{p_m}$ is equivalent to one of the form $U^{m,\lambda} = \mathrm{Ind}_{\overline{G}_{p_m}}^{\overline{G}}(D^{m,\lambda})$ where $D^{m,\lambda}$ is the representation of $\overline{G}_{p_m}$

$$(\boldsymbol{a}, b, c, \boldsymbol{0}, r, x) \mapsto e^{i(mc+\lambda x)}.$$

Moreover, the set $\{U^{m,\lambda} \ : \ m, \lambda \in \mathbb{R}, \ m > 0\}$ is a family of physically inequivalent representations of $\overline{G}$.

To compute explicitly $U^{m,\lambda}$, we observe that the orbit

$$\overline{G}[p_m] = \{(\boldsymbol{p}, E, m) \ : \ E - \frac{\boldsymbol{p}^2}{2m} = 0\}$$

can be identified with $\mathbb{P}^2$ using the map

$$\mathbb{P}^2 \ni \boldsymbol{p} \longleftrightarrow \left(\boldsymbol{p}, \frac{\boldsymbol{p}^2}{2m}, m\right) \in \overline{G}[p_m].$$

With respect to this identification the action of $\overline{G}$ on the orbit becomes

$$(\boldsymbol{a}, b, c, \boldsymbol{v}, r, x)[\boldsymbol{p}] = \delta(r)\boldsymbol{p} + m\boldsymbol{v}$$

so that the Lebesgue measure $d\boldsymbol{p}$ on $\mathbb{P}^2$ is $\overline{G}$-invariant. We consider the section

$$\beta : \mathbb{P}^2 \to \overline{G}, \ \boldsymbol{p} \mapsto \left(\boldsymbol{0}, 0, 0, \frac{\boldsymbol{p}}{m}, 0, 0\right)$$

for the action of $\overline{G}$ on $\mathbb{P}^2$. The representation $U^{m,\lambda}$ of $\overline{G}$ acts in $L^2(\mathbb{P}^2, d\boldsymbol{p})$ as

$$\left(U^{m,\lambda}_{(\boldsymbol{a},b,c,\boldsymbol{v},r,x)} f\right)(\boldsymbol{p}) = e^{i\left(\frac{b}{2m}\boldsymbol{p}^2 - \boldsymbol{p}\cdot\boldsymbol{a} + mc\right)} e^{i\lambda\left(x + \frac{1}{m}(v_1 p_2 - v_2 p_1)\right)} f(\delta(-r)(\boldsymbol{p} - m\boldsymbol{v})).$$

From the explicit form of $U^{m,\lambda}$ one readily gets that the angular momentum, i.e. the selfadjoint operator that generates the 1-parameter subgroup of rotations, has only the *orbital* part, so that the elementary particles in $2+1$ dimensions have no *spin*. However, they acquire a new *charge* λ, which is not of a space-time origin, but arises from the structure of the multipliers. If $\lambda \neq 0$, the two linear momenta do not commute.

We add some final comments.

1. The characters of G^* are

$$G^* \ni (\boldsymbol{a}, b, \boldsymbol{v}, r) \mapsto e^{iEb} e^{i\mu r} \in \mathbb{T},$$

where $E, \mu \in \mathbb{R}$. The set V of characters of K that extend to G^* is

$$V = \{(c, \boldsymbol{0}, 0, \boldsymbol{0}, 2\pi n, x) \mapsto z^n \ : \ z \in \mathbb{T}\} \simeq \mathbb{T}.$$

The group V is a closed subgroup of $\widehat{K} = \mathbb{P}^2 \times \mathbb{T}$ and $K_0 = \mathbb{R}^2$. Applying Corollary 5, $H^2(G, \mathbb{T})$ is isomorphic to $\mathbb{R}^2$. In particular, any $\mathbb{T}$-multiplier of G is equivalent to one of the form

$$\left((\boldsymbol{a}, b, \boldsymbol{v}, R), (\boldsymbol{a}', b', \boldsymbol{v}', R') \right) \mapsto e^{im(\boldsymbol{v} \cdot R\boldsymbol{a}' + \frac{1}{2} b' \boldsymbol{v}^2)} e^{i\lambda R(v_1 v'_2 - v_2 v'_1)} ,$$

where $m, \lambda \in \mathbb{R}^2$, $\boldsymbol{v} = (v_1 . v_2)$ and $\boldsymbol{v}' = (v'_1 . v'_2)$.

2. From the explicit form of the characters of G^* one has that, for all $E, \mu \in \mathbb{R}$, the representation

$$(\boldsymbol{a}, b, c, \boldsymbol{v}, r, x) \mapsto e^{i(Eb + \mu r)} U^{m,\lambda}_{(\boldsymbol{a}, b, c, \boldsymbol{v}, r, x)}$$

is physically equivalent to $U^{m,\lambda}$. Hence the angular momentum and the energy are both defined up to an additive constant. For the energy this phenomenon is well known in $3 + 1$ dimensions, while it does not occur for the angular momentum.
3. The admissibility condition (3.6) gives rise to two superselection rules that do not allow superpositions of states with different *mass* m or with different *charge* λ. However, there is no superselection rule connected with the *spin*.

6 Galilei Invariant Wave Equations

According to the results of the previous sections, a free elementary quantum object that is invariant under the Galilei group is described by an irreducible unitary representation of the corresponding universal central extension. The representation acts on a Hilbert space of square integrable functions over the momentum space, so that the vector states of the system are functions of the linear momentum and energy. This description is *unnatural* when the object is interacting with an external field, which is a function of space-time variables. In such a case one would like to characterise the states of the system as solutions of *wave equations*, which are partial differential equations in the space-time variables. Usually these equations are deduced from a Lagrangian which is invariant with respect to a suitable representation of the symmetry group and the states of the system are regarded as classical (differentiable) fields. However, in doing so one hides the Hilbert space structure of the theory and the mathematical problems connected with the fact that state vectors are, in general, only square integrable and not necessarily differentiable functions.

To overcome this problem, in Sect. 6.1 we describe vector states as vector valued distributions on the space-time that are weak solutions of invariant *local* operators, called wave operators. These wave operators are characterised by Theorem 5 below and we use this result to describe the Galilei invariant wave equations both in $3+1$ and in $2+1$ dimensions.

Section 6.2 is devoted to studying the $3+1$ dimensional case. For each elementary particle we find two classes of wave equations. The first one is the usual Schrödinger equation, which is a second order differential operator and which is not invariant with respect to the Galilei group, but only with respect to its universal central extension. This fact reflects the true projective character of unitary representations associated with Galilei invariant objects.

The main feature of the second class is that the wave operator is a differential operator of the first order in the space-time variables as is the Dirac equation for the relativistic electron. If the interaction with an electromagnetic field is introduced on this latter wave equation by means of the minimal coupling principle, the particle acquires an internal magnetic momentum with gyromagnetic ratio $g = \frac{1}{j}$, where j is the spin of the particle. We notice that, if one introduces the interaction on the Schrödinger equation, one does not obtain any coupling between spin and magnetic field.

G. Cassinelli, E. De Vito, P.J. Lahti, and A. Levrero, *The Theory of Symmetry Actions in Quantum Mechanics*, Lect. Notes Phys. **654**, pp. 73–87
http://www.springerlink.com/

Section 6.3 describes the $2+1$ dimensional wave equations. In this case, there are two kinds of elementary particles that can describe physical systems. The first one does not admit wave equations at all, whereas, for each particle of the second class, there is essentially only one wave equation of first order and such that the vector space where the distributions take values has minimal dimension. The particles described by this wave equation interact with an electromagnetic field as three-dimensional spin-1/2 particles.

6.1 Wave Equations

In this section we briefly recall the main results on the wave equations, following the theory developed in [10].

To begin with we recall some basic definitions concerning distributions, see [21] for a full account. Given $n \in \mathbb{N}$, let $\mathcal{D}(\mathbb{R}^n)$ be the space of C^∞ functions $f : \mathbb{R}^n \to \mathbb{C}$ with compact support and $\mathcal{D}'(\mathbb{R}^n)$ the corresponding space of distributions, that is, the space of (continuous) linear functionals

$$T : \mathcal{D}(\mathbb{R}^n) \to \mathbb{C}, \;\; f \mapsto \langle T, f \rangle.$$

We recall that any continuous function $\phi : \mathbb{R}^n \to \mathbb{C}$ defines a distribution T_ϕ as

$$\langle T_\phi, f \rangle = \int_{\mathbb{R}^n} \phi(x) f(x)\, dx,$$

where dx is the Lebesgue measure on $\mathbb{R}^n$. In the following we identify T_ϕ with ϕ. Finally, the support of a distribution T is defined as the smallest closed set $\operatorname{supp} T \subseteq \mathbb{R}^n$ having the following property: for all $f \in \mathcal{D}(\mathbb{R}^n)$, if $\operatorname{supp} f \cap \operatorname{supp} T = \emptyset$, then $\langle T, f \rangle = 0$.

In order to take care of the spin of the object we have to consider vector valued distributions. For any $m \in \mathbb{N}$, we let $\mathcal{D}(\mathbb{R}^n)^m$ and $\mathcal{D}'(\mathbb{R}^n)^m$ be the m-fold Cartesian products of $\mathcal{D}(\mathbb{R}^n)$ and $\mathcal{D}'(\mathbb{R}^n)$, respectively, and we write

$$\langle T, f \rangle := \sum_{i=1}^{m} \langle T_i, f_i \rangle,$$

where $T = (T_1, \dots, T_m) \in \mathcal{D}'(\mathbb{R}^n)^m$ and $f = (f_1, \dots, f_m) \in \mathcal{D}(\mathbb{R}^n)^m$.

To give the definition of an invariant wave equation we replace, as usual, the Galilei group G by its universal central extension $\overline{G}$. The group $\overline{G}$ is a semidirect product $A \times' H$, where H is a Lie group and the normal Abelian factor A is $\mathbb{R}^n$, with $n = 5$ in $3+1$ dimensions and $n = 4$ in the $2+1$ dimensional case.

It is natural to assume that spin is invariant with respect to translations and that it transforms according to a representation L with respect to the action of the homogeneous part. Hence, we fix a finite dimensional representation L of H acting on $\mathbb{C}^m$.

To simplify the notation, we extend L to a representation of $\overline{G}$ in a trivial way,

$$L(a,h) := L(h) , \qquad a \in \mathbb{R}^n,\ h \in H.$$

We use it to define a *geometric* action of $\overline{G}$ on $\mathcal{D}'(\mathbb{R}^n)^m$ in the following way: given $\overline{g} \in \overline{G}$, for all $T \in \mathcal{D}'(\mathbb{R}^n)^m$, define $\Lambda_{\overline{g}}T$ as the distribution

$$\langle \Lambda_{\overline{g}}T, f\rangle := \langle T, f^{\overline{g}}\rangle , \qquad f \in \mathcal{D}(\mathbb{R}^n)^m,$$

where, for all $f \in \mathcal{D}(\mathbb{R}^n)^m$,

$$(f^{\overline{g}})(x) := L(\overline{g})^t f(\overline{g}[x]) , \qquad x \in \mathbb{R}^n,$$

and $L(\overline{g})^t$ is the transpose of the matrix $L(\overline{g})$.

One can check that $\Lambda_{\overline{g}}T$ is well defined and $\overline{g} \mapsto \Lambda_{\overline{g}}$ is a (differentiable) representation of $\overline{G}$ acting on $\mathcal{D}'(\mathbb{R}^n)^m$, see, for example, [34] for definitions.

Since $\mathbb{C}^m$ is a complex vector space, we can also consider the adjoint representation L^*, that is,

$$L^*(\overline{g}) := L(\overline{g}^{-1})^* , \qquad \overline{g} \in \overline{G},$$

where $L(\overline{g}^{-1})^*$ is the adjoint (transpose and complex conjugate) of $L(\overline{g}^{-1})$. As above, we let

$$\langle \Lambda^*_{\overline{g}}T, f\rangle := \langle T, f^{\overline{g}^*}\rangle , \qquad x \in \mathbb{R}^n,$$

where

$$(f^{\overline{g}^*})(x) := (L^*)(\overline{g})^t f(\overline{g}[x]) = \overline{L(\overline{g}^{-1})} f(g[x]) , \qquad x \in \mathbb{R}^n.$$

The use of the transpose matrix is motived by the following fact. If $T \in \mathcal{D}'(\mathbb{R}^n)^m$ is the function $\phi : \mathbb{R}^n \to \mathbb{C}^m$ and $\overline{g} \in \overline{G}$, then $\Lambda_{\overline{g}}T$ is the function

$$(\Lambda_{\overline{g}}T)(x) = L(\overline{g})\phi(\overline{g}^{-1}[x]) , \qquad x \in \mathbb{R}^n,$$

and $\Lambda^*_{\overline{g}}T$ is the function

$$(\Lambda^*_{\overline{g}}T)(x) = L^*(\overline{g})\phi(\overline{g}^{-1}[x]) , \qquad x \in \mathbb{R}^n.$$

With the above background the definition of an invariant wave equation can now be given.

Definition 16. An *L-invariant wave equation* is a family $(D_i)_{i=1}^N$ of continuous operators on $\mathcal{D}'(\mathbb{R}^n)^m$ satisfying the following conditions.

i) For any $T \in \mathcal{D}'(\mathbb{R}^n)^m$ and $i = 1, \ldots , N$,

$$\operatorname{supp} D_i T \subset \operatorname{supp} T. \tag{6.1}$$

ii) For each $i = 1, \ldots, N$, one of the following two conditions holds:

$$\Lambda_{\overline{g}}(D_i T) = D_i(\Lambda_{\overline{g}} T)\,, \qquad \overline{g} \in \overline{G},\ T \in \mathcal{D}'(\mathbb{R}^n)^m, \tag{6.2}$$

$$\Lambda^*_{\overline{g}}(D_i T) = D_i(\Lambda_{\overline{g}} T)\,, \qquad \overline{g} \in \overline{G},\ T \in \mathcal{D}'(\mathbb{R}^n)^m. \tag{6.3}$$

iii) The vector space of weak solutions of the system

$$\begin{cases} D_1 T &= 0, \\ &\cdots \\ D_N T &= 0, \end{cases} \tag{6.4}$$

contains a unique subspace $\mathcal{H}$ such that $\mathcal{H}$ is a Hilbert space with respect to a suitable sesquilinear form $\langle \cdot, \cdot \rangle_{\mathcal{H}}$, inducing a topology finer than the original one, and the action of Λ on $\mathcal{H}$ is invariant, unitary and irreducible.

A continuous operator on $\mathcal{D}'(\mathbb{R}^n)^m$ that decreases supports, that is, has the property (6.1), and which satisfies condition (6.2) (or condition (6.3)) is called an *invariant wave operator* (or **-invariant wave operator*).

The above definition is motived by the following observations. The need to define wave equations in terms of N wave operators is in order to take care of the fact that the group A is bigger than the physical space-time $\mathbb{R}^4$ (or $\mathbb{R}^3$). Condition (6.1) is the requirement that the physical system be described by *local* dynamical laws. Conditions (6.2) and (6.3) assure that the kernels of the operators D_i are $\overline{G}$-invariant. In particular, the first one is *natural* from a geometrical point of view and the second one implies that the *formal* Lagrangian functions

$$\mathcal{L}_i(T) = \sum_{j=1}^{m} \int_{\mathbb{R}^n} T_j(x)(D_i T)_j(x)\, dx$$

are $\overline{G}$-invariant. Finally, the system of (6.4) singles out exactly one elementary system, namely the one described by the unitary irreducible representation U, where U denotes the restriction of Λ to $\mathcal{H}$. In the following, we say that the wave equation defined by the system (6.4) is *associated* with U.

Now we fix a G-elementary particle, described by a unitary irreducible representation U of $\overline{G}$ and we are searching for wave equations associated with U. The following theorem gives a classification. To state the result we wish to recall some facts. Since A is the vector space $\mathbb{R}^n$, its dual is $\mathbb{P}^n = \mathbb{R}^n$.

The dual space $\mathbb{P}^n$ of $\mathbb{R}^n$ can be split into a disjoint union of orbits of a family of points $\{p_s\}_{s\in I} \subset \mathbb{P}^n$, that is, $\mathbb{P}^n = \cup_{s\in I} \overline{G}[p_s]$, where, for $s \neq t$, $\overline{G}[p_s] \cap \overline{G}[p_s] = \emptyset$. Moreover, every unitary irreducible representation of $\overline{G}$ is equivalent to one of the form $\mathrm{Ind}_{\overline{G}_{p_s}}^{\overline{G}}(p_s \otimes \pi)$, where π is an irreducible unitary representation of $H_{p_s} = \overline{G}_{p_s} \cap H$ acting on $\mathcal{K}_\pi$. We denote by $\mathrm{Hom}_{H_{p_s}}(\mathcal{K}_\pi; \mathbb{C}^m)$ the vector space of continuous operators from $\mathcal{K}_\pi$ to $\mathbb{C}^m$ that intertwine the representations π and L, viewed as a representation of H_{p_s}.

The representation U is thus of the form $U = \mathrm{Ind}_{\overline{G}_{p_0}}^{\overline{G}}(p_0 \otimes \tau)$ for some $p_0 \in \mathbb{P}^n$ and some irreducible unitary representation τ of $H_{p_0} = \overline{G}_{p_0} \cap H$.

Theorem 5. *For any $p \in \mathbb{P}^n$, let $(M_1(p), \dots, M_N(p))$ be complex $m \times m$ matrices satisfying the following conditions:*

1. *for all $i = 1, \dots, N$, the map $p \mapsto M_i(p)$ is polynomial;*
2. *for each $i = 1, \dots, N$, one of the following two conditions is satisfied,*

$$L(h)M_i(h^{-1}[p])L(h^{-1}) = M_i(p) \tag{6.5}$$

$$L(h)^* M_i(h^{-1}[p])L(h^{-1}) = M_i(p) \tag{6.6}$$

   for all $h \in H$ and $p \in \mathbb{P}^n$;

3. *there is a unique, up to a constant, $B \in \mathrm{Hom}_{H_{p_0}}(\mathcal{K}_\tau; \mathbb{C}^m)$, $B \neq 0$, such that*

$$M_i(p_0)B = 0$$

   for all $i = 1, \dots, N$;

4. *for any irreducible representation $\mathrm{Ind}_{\overline{G}_{p_s}}^{\overline{G}}(p_s \otimes \pi)$ not unitarily equivalent to U and for all $B \in \mathrm{Hom}_{H_{p_s}}(\mathcal{K}_\pi; \mathbb{C}^m)$, $B \neq 0$, there exists an $i = 1, \dots, N$, such that*

$$M_i(p_s)B \neq 0.$$

For all $i = 1, \dots, N$, define the operators D_i as

$$\langle D_i T, f \rangle := \langle T, \mathcal{F}^{-1} M_i^t \mathcal{F} f \rangle, \tag{6.7}$$

where $T \in \mathcal{D}'(\mathbb{R}^n)^m$, $f \in \mathcal{D}(\mathbb{R}^n)^m$ and $\mathcal{F}$ is the Fourier transform. The operators $(D_1, \dots, D_N)$ are L-invariant wave equations associated with U.

The above theorem holds under two technical assumptions, which are always satisfied for the Galilei groups. First of all, the $\overline{G}$-invariant measure ν on the orbit $\overline{G}[p_0]$ defines a *tempered* distribution T_ν by means of

$$\langle T_\nu, f \rangle := \int_{\overline{G}[p_0]} f(x)\, d\nu(x),$$

see [21] for definitions. Moreover the representation L is *at most of polynomial growth on* $\overline{G}[p_0]$, that is,

$$\|L(\beta(p))\| \leq C(1 + |p|^k)\,, \qquad p \in \overline{G}[p_0],$$

where $C > 0$ and $k \in \mathbb{N}$ are suitable constants, $|p|$ is the Euclidean norm of p, $\|L(\beta(p))\|$ is the usual matrix norm and β is a section from $\overline{G}[p_0]$ to H.

The proof of this result uses in a deep way the theory of quasi-invariant distributions [7]. The reader may wish to consult [10] for a proof. Here we stress only the following facts.

Remark 5. The condition that $\mathrm{Ind}_{\overline{G}_{p_s}}^{\overline{G}}(p_s \otimes \pi)$ is not unitarily equivalent to U implies that either $p_s \neq p_0$ or, if $p_s = p_0$, π is not equivalent to τ.

Remark 6. Equation (6.5) implies that, for all $s \in I$, $M(p_s) \in \mathrm{Hom}_{H_{p_s}}(\mathbb{C}^m; \mathbb{C}^m)$ and (6.6) $M(p_s) \in \mathrm{Hom}_{H_p}(\mathbb{C}^m; (\mathbb{C}^m)^*)$, where $(\mathbb{C}^m)^*$ denotes the vector space $\mathbb{C}^m$ endowed with the representation adjoint to L. In both cases, $M(p_s)$ defines uniquely $M(p)$ on the corresponding orbit $\overline{G}[p_s]$.

The condition in item (3) of Theorem 5 implies that the space $\mathrm{Hom}_{H_{p_0}}(\mathcal{K}_\tau; \mathbb{C}^m)$ is non zero and this is equivalent to the fact that L restricted to H_{p_0} contains τ as a subrepresentation. In the same way, the condition in item (4) has to be checked only for representations π that are contained in L.

We close this section with some further observations. Since the unitary equivalence is stronger than the physical equivalence, the equivalence class of U is not uniquely defined by the quantum system and, hence, we can obtain different wave equations associated with the same elementary particle. This is not surprising since, for instance, the Hamiltonian operator is defined only up to a constant and hence also the Schrödinger equation. Obviously, the physical content of the theory has to be the same in all those cases.

Due to the fact that the matrices $M_i(p)$ are polynomial in p, $\mathcal{F}^{-1}M^t\mathcal{F}f$ is in $\mathcal{D}(\mathbb{R}^n)^m$ for all $f \in \mathcal{D}(\mathbb{R}^n)^m$, so that (6.7) is well defined and it implies that the wave operators D_i are finite order differential operators with constant coefficients, that is,

$$D_i = \sum_\nu C_\nu \frac{\partial}{\partial^{\nu_1} x_1 \ldots \partial^{\nu_n} x_n},$$

where $\nu = (\nu_1, \ldots, \nu_n)$ is an integer multi-index and C_ν are *constant* $m \times m$ complex matrices. Moreover, since D_i are differential operators, one can also prove [10] that the vector states of the elementary particle, which are solutions of the wave equation, are in fact tempered distributions. Finally, the degree of the polynomial $M_i(p)$ defines the order of the differential operator D_i. It is always possible to reduce the order of the corresponding differential equation $D_i T = 0$ by adding auxiliary degrees of freedom. Nevertheless, in general, one obtains wave operators that do not satisfy (6.2) or (6.3). Therefore, it is of interest to find representations L of H admitting a wave equation given by *first order* differential operators. In this case we say that the wave equation is of *Dirac type*.

In [10], the following stronger result is proved. If the two technical assumptions mentioned after Theorem 5 hold for every orbit $\overline{G}[p_s]$, $s \in I$, then every L-invariant wave equation $(D_1, \ldots, D_N)$ is of the form given by Theorem 5.

6.2 The 3 + 1 Dimensional Case

From Chap. 5 we know that the free elementary Galilei objects in 3 + 1 dimensions having physical meaning are described by mass and spin. Thus we fix the mass $m_0 > 0$ and the spin j with $2j \in \mathbb{N}$ and we search for the wave equations associated with $U^{(m_0,j)} = \mathrm{Ind}_{\overline{G}_{p_0}}^{\overline{G}}(p_0 \otimes \mathbb{D}^j)$, where $p_0 = (\mathbf{0}, 0, m_0)$ and $\mathbb{D}^j$ is the representation of $SU(2)$ acting on $\mathbb{C}^{2j+1}$, see Sect. 5.2.3 of Chap. 5 for the notations.

We define the Fourier transform between $A = \mathbb{R}^5$ and $\hat{A} = \mathbb{P}^5$ in a such way that it gives the following correspondence between the multiplicative operators on $\mathcal{D}'(\mathbb{P}^5)$ and the differential operators on $\mathcal{D}'(\mathbb{R}^5)$:

$$\begin{aligned} \boldsymbol{p} &\leftrightarrow i\boldsymbol{\nabla} \\ E &\leftrightarrow -i\frac{\partial}{\partial t} \\ c &\leftrightarrow -i\frac{\partial}{\partial \xi}. \end{aligned}$$

The first step is to select a finite-dimensional representation L of H. In the final section of this chapter we shall briefly present a method to obtain finite-dimensional representations of H. To choose among them, we can consider two different kinds of constraints. The first one is that the vector space carrying the representation L has a minimal dimension. On the other hand, we can require the wave equation to be of the Dirac type.

We consider the first case. Let $\mathbb{D}^j$ be the irreducible representation of $SU(2)$ acting on $\mathbb{C}^{2j+1}$, and extend it to a representation L^j of H in a trivial way, that is, $L^j(\boldsymbol{v}, h) = \mathbb{D}^j(h)$. Define the mappings M_1 and M_2 from $\mathbb{P}^5$ into the space of $(2j+1) \times (2j+1)$ matrices as

$$\begin{aligned} M_1(\boldsymbol{p}, E, m) &= (m - m_0) I_{2j+1} \\ M_2(\boldsymbol{p}, E, m) &= (2mE - \boldsymbol{p}^2) I_{2j+1}. \end{aligned}$$

One can easily check that M_1 and M_2 satisfy all the conditions of Theorem 5. Using relation (6.7), one obtains the explicit form of the corresponding wave operators (D_1, D_2):

$$\begin{aligned} D_1 &= (i\frac{\partial}{\partial \xi} - m_0) I_{2j+1} , \\ D_2 &= \left(-2\frac{\partial^2}{\partial \xi \partial t} + \boldsymbol{\nabla}^2\right) I_{2j+1}. \end{aligned}$$

To show that the equations are the usual Schrödinger equation, consider a function $\phi \in C^\infty(\mathbb{R}^5, \mathbb{C}^{2j+1})$ such that

$$\begin{cases} D_1\phi = 0 \\ D_2\phi = 0 \end{cases} ,$$

in the sense of a distribution. From the first equation it follows that

$$\phi(\boldsymbol{x}, t, \xi) = e^{-im_0\xi}\psi(\boldsymbol{x}, t) ,$$

where ψ is in $C^\infty(\mathbb{R}^4, \mathbb{C}^{2j+1})$ and from the second equation one gets that ψ satisfies the Schrödinger equation

$$i\frac{\partial \psi}{\partial t}(\boldsymbol{x}, t) = -\frac{1}{2m_0}\boldsymbol{\nabla}^2\psi(\boldsymbol{x}, t).$$

One has to consider two differential operators in order to assure the correct invariant property of the wave equation (it is well known that the Schrödinger equation is not Galilei invariant). Moreover it is clear that the dimension of $\mathbb{C}^{2j+1}$ is minimal.

Now we address the problem of finding a wave equation of the Dirac type. We consider first the case $j > \frac{1}{2}$. Let L^j be the representation of H, acting on

$$V^j = \mathbb{C}^{2j-1} \oplus \mathbb{C}^{2j+1} \oplus \mathbb{C}^{2j+1},$$

given by

$$L^j(\boldsymbol{v}, h) = \begin{pmatrix} \mathbb{D}^{j-1}(h) & 0 & \boldsymbol{v}\cdot \boldsymbol{T}^{(j-1)j}\mathbb{D}^j(h) \\ 0 & \mathbb{D}^j(h) & \boldsymbol{v}\cdot \boldsymbol{H}^j\mathbb{D}^j(h) \\ 0 & 0 & \mathbb{D}^j(h) \end{pmatrix}, \tag{6.8}$$

with $\boldsymbol{v} \in \mathcal{V}$, $h \in SU(2)$ and the matrices $T^{(j-1)j}$ and H^j are introduced in Sect. 6.4 below. We notice that $\dim V^j = 6j + 1$.

Let D_1 and D_2 be the differential operators

$$D_1 = (i\frac{\partial}{\partial \xi} - m_0)I_{6j+1} \ ,$$

$$D_2 = i\begin{pmatrix} I_{2j-1}\frac{\partial}{\partial \xi} & 0 & \boldsymbol{T}^{(j-1)j}\cdot \boldsymbol{\nabla} \\ 0 & I_{2j+1}\frac{\partial}{\partial \xi} & \boldsymbol{H}^j\cdot \boldsymbol{\nabla} \\ (\boldsymbol{T}^{(j-1)j})^*\cdot \boldsymbol{\nabla} & \boldsymbol{H}^j\cdot \boldsymbol{\nabla} & 2j^2 I_{2j+1}\frac{\partial}{\partial t}\cdot \end{pmatrix}.$$

The above wave equation coincides, at least formally, with the one found by Hurley with Lagrangian methods, [22]. Applying Theorem 5 one has the following result.

Proposition 11. *The set (D_1, D_2) is a Dirac type L^j-invariant wave equation associated with the Galilean elementary particle of mass m_0 and spin j.*

Proof. Using the notation of Theorem 5, we have that $p_0 = (\boldsymbol{0}, 0, m_0)$, $\overline{G}_{p_0} \cap H = SU(2)$, $\tau = \mathbb{D}^j$ and $\mathcal{K}_\tau = \mathbb{C}^{2j+1}$. Moreover, the orbit $\overline{G}[p_0]$ is isomorphic to $\mathbb{P}^3$ and, under this isomorphism, the invariant measure ν on $\overline{G}[p_0]$ is the Lebesgue measure that clearly defines a tempered distribution. Finally, the representation L^j is at most of polynomial growth on the orbit $\overline{G}[p_0]$. Indeed, using the section

$$\overline{G}[p_0] \ni (\boldsymbol{p}, E, m) \mapsto (\frac{\boldsymbol{p}}{m_0}, e) \in H \ ,$$

one has that $\left\| L^j(\frac{p}{m_0}, e)\right\| \leq (1 + C|\boldsymbol{p}|^2)$, where C is a suitable constant. Theorem 5 can thus be applied.

Define two functions M_1 and M_2 from $\mathbb{P}^5$ to the space of $(6j+1)\times(6j+1)$ matrices as

$$M_1(\boldsymbol{p}, E, m) = (m - m_0)I_{6j+1} ,$$

$$M_2(\boldsymbol{p}, E, m) = \begin{pmatrix} mI_{2j-1} & 0 & -\boldsymbol{p}\cdot\boldsymbol{T}^{(j-1)j} \\ 0 & mI_{2j+1} & -\boldsymbol{p}\cdot\boldsymbol{H}^j \\ -\boldsymbol{p}\cdot(\boldsymbol{T}^{(j-1)j})^* & -\boldsymbol{p}\cdot\boldsymbol{H}^j & 2j^2EI_{2j+1}. \end{pmatrix} .$$

One can easily check that, for all $T \in \mathcal{D}'(\mathbb{R}^5)^{6j+1}$ and $f \in \mathcal{D}(\mathbb{R}^5)^{6j+1}$,

$$\langle D_i T, f\rangle = \langle T, \mathcal{F}^{-1}M_i^t\mathcal{F}f\rangle, \qquad i = 1, 2.$$

Obviously the maps M_1 and M_2 are polynomials with maximal degree 1 and M_1 satisfies (6.5). We now show that M_2 satisfies (6.6). Indeed, it is enough to verify this on the set

$$\Omega = \{(\boldsymbol{p}, E, m) \in \mathbb{P}^5 \ : \ m \neq 0\},$$

which is dense in $\mathbb{P}^5$. By means of an explicit computation one has that for all $(\boldsymbol{p}, E, m) \in \Omega$,

$$(\boldsymbol{p}, E, m) = (\frac{\boldsymbol{p}}{m}, e)[(m, \boldsymbol{0}, E - \frac{\boldsymbol{p}^2}{2m})].$$

Moreover, define for all $a, b \in \mathbb{R}$ the matrix

$$\Delta(a, b) := \begin{pmatrix} aI_{2j-1} & 0 & 0 \\ 0 & aI_{2j+1} & 0 \\ 0 & 0 & 2j^2bI_{2j+1} \end{pmatrix}.$$

The matrix $\Delta(a, b)$ is in the commuting ring of the restriction of L^j to $SU(2)$ (observe that if $h \in SU(2)$, then $(L^j)^*(h) = L^j(h)$). Finally, for all $(\boldsymbol{p}, E, m) \in \Omega$,

$$M_2(\boldsymbol{p}, E, m) = (L^j)^*((\frac{\boldsymbol{p}}{m}, e))\Delta(m, E - \frac{\boldsymbol{p}^2}{2m})(L^j)((-\frac{\boldsymbol{p}}{m}, e)),$$

where we use (6.12) below to linearise the term proportional to $\frac{\boldsymbol{p}^2}{2m}$.

We now show that the conditions (3) of Theorem 5 are satisfied. Indeed, let $B \in \mathrm{Hom}_{SU(2)}(\mathbb{C}^{2j+1}; \mathbb{C}^{6j+1})$. Then B is of the form

$$B(w) = (0, \beta w, \alpha w) \in \mathbb{C}^{2j-1} \oplus \mathbb{C}^{2j+1} \oplus \mathbb{C}^{2j+1},$$

where $\alpha, \beta \in \mathbb{C}$. Since $M_1(p_0) = 0$, obviously $M_1(p_0)B = 0$, whereas, by an explicit computation, the condition $M_2(p_0)B = 0$ is equivalent to $\beta = 0$. Hence, there is a unique $B \neq 0$, up to a constant, satisfying condition (3).

Consider now condition (4). Let $p_s = (\mathbf{0}, \varepsilon, m) \in \mathbb{P}^5$ and π an irreducible unitary representation of $\overline{G}_{p_s} \cap H = SU(2)$ such that

$$\mathrm{Ind}_{\overline{G}_{p_s}}^{\overline{G}}(p \otimes \pi) \text{ is not equivalent to } \mathrm{Ind}_{\mathbb{R}^5 \times' SU(2)}(p_0 \otimes \mathbb{D}^j), \tag{6.9}$$

then, taking into account Remark 5,

1. if $m \neq m_0$, then $M_1(p_s) = (m - m_0)I_{6j+1}$ is clearly invertible, so that if $B \in \mathrm{Hom}_{H_{p_s}}(\mathcal{K}_\pi; \mathbb{C}^{6j+1})$, $B \neq 0$, then $M_1(p_s)B \neq 0$;
2. if $m = m_0$, then
   a) if $\varepsilon \neq 0$,

$$M_2(p_s) = \begin{pmatrix} m_0 I_{2j-1} & 0 & 0 \\ 0 & m_0 I_{2j+1} & 0 \\ 0 & 0 & 2j^2 \varepsilon I_{2j+1}, \end{pmatrix},$$

   which is invertible, and, as above, we can conclude that the relation (4) holds with $i = 2$;

   b) if $\varepsilon = 0$, then, by (6.9), $\pi = \mathbb{D}^{j'}$ with $j' \neq j$. Moreover, the fact that

$$\mathrm{Hom}_{H_{SU(2)}}(\mathcal{K}_{p_s}; \mathbb{C}^{6j+1}) \neq \{0\},$$

   implies that $j' = j - 1$ and that the range of B is contained in the subspace $\mathbb{C}^{2j-1} \oplus \{0\} \oplus \{0\}$. Hence $M_2(p_s)B \neq 0$ if $B \neq 0$. □

For $j = \frac{1}{2}$ the vector space $\mathbb{C}^{2j-1}$ reduces to the zero vector and the matrices $\boldsymbol{T}^{(j-1)j}$ are the zero matrices, so that the wave operator D_2 is given by

$$D_2 = i \begin{pmatrix} I_2 \frac{\partial}{\partial \xi} & \boldsymbol{H}^{\frac{1}{2}} \cdot \boldsymbol{\nabla} \\ \boldsymbol{H}^{\frac{1}{2}} \cdot \boldsymbol{\nabla} & \frac{1}{2} I_2 \frac{\partial}{\partial t} \end{pmatrix}.$$

This was found for the first time by Lévy-Leblond, [24]. The main difference with the case $j \geq 1$ is that D_2 is both *-invariant and invariant. From the mathematical point of view this phenomenon is a consequence of the fact that the matrices $\boldsymbol{H}^{\frac{1}{2}}$, which are proportional to the Pauli matrices, satisfy also the anticommutation relations

$$H_i^{\frac{1}{2}} H_j^{\frac{1}{2}} + H_j^{\frac{1}{2}} H_i^{\frac{1}{2}} = \frac{1}{2}\delta_{ij}.$$

The case $j = 0$ requires small modifications. Let L^0 be the representation of H acting on $\mathbb{C}^3 \oplus \mathbb{C}$ given by

$$L^0(v, h) = \begin{pmatrix} \delta(h) & \boldsymbol{v} \\ 0 & 1 \end{pmatrix},$$

where $v \in \mathcal{V}$ and $h \in SU(2)$. Let D_1 and D_2 be the differential operators

$$D_1 = (i\frac{\partial}{\partial \xi} - m_0)I_4 ,$$

$$D_2 = i \begin{pmatrix} I_3 \frac{\partial}{\partial \xi} & \boldsymbol{\nabla} \\ \boldsymbol{\nabla} & 2\frac{\partial}{\partial t} \end{pmatrix} .$$

The proof that (D_1, D_2) is a Dirac type wave equation for the spinless particle of mass m_0 is the same as the one of Proposition 11. Observe that in this case D_2 is a *-invariant wave operator and the dimension of the vector space carrying the representation is 4.

6.2.1 The Gyromagnetic Ratio

In the previous section we proved that with each elementary particle of mass m_0 and spin j there is associated an L-invariant Dirac type wave equation (D_1, D_2). We now introduce the interaction of the particle with a given external electromagnetic field $(\boldsymbol{E}, \boldsymbol{B})$. We stress that $(\boldsymbol{E}, \boldsymbol{B})$ are not dynamical variables and, obviously, can destroy the Galilei invariance. The interaction will be introduced on the *free* wave equation by means of the minimal coupling principle. Taking into account that the variable ξ has no physical meaning the prescription of the minimal coupling is

$$\begin{aligned} -i\boldsymbol{\nabla} &\mapsto -i\boldsymbol{\mathcal{D}} := -i\boldsymbol{\nabla} - q\boldsymbol{A}(\boldsymbol{x}, t) \\ i\frac{\partial}{\partial t} &\mapsto i\mathcal{D}_4 := i\frac{\partial}{\partial t} - qV(\boldsymbol{x}, t) \\ i\frac{\partial}{\partial \xi} &\mapsto i\mathcal{D}_5 := i\frac{\partial}{\partial \xi} , \end{aligned}$$

where V and $\boldsymbol{A}$ are potentials defined by $(\boldsymbol{E}, \boldsymbol{B})$ and q is the electric charge. In doing so, for $j > 0$, the wave equation becomes

$$D_1^{int} = (i\mathcal{D}_5 - m_0)I_{6j+1} ,$$

$$D_2^{int} = \begin{pmatrix} iI_{2j-1}\mathcal{D}_5 & 0 & i\boldsymbol{T}^{(j-1)j} \cdot \boldsymbol{\mathcal{D}} \\ 0 & iI_{2j+1}\mathcal{D}_5 & i\boldsymbol{H}^j \cdot \boldsymbol{\mathcal{D}} \\ (\boldsymbol{T}^{(j-1)j})^* \cdot \boldsymbol{\mathcal{D}} & i\boldsymbol{H}^j \cdot \boldsymbol{\mathcal{D}} & 2j^2 iI_{2j+1}\mathcal{D}_4 . \end{pmatrix} .$$

To explain the meaning of this equation, suppose that Φ is a smooth function from $\mathbb{R}^5$ to $\mathbb{C}^{6j+1}$ satisfying

$$D_1^{int}\Phi = 0 \qquad \text{and} \qquad D_2^{int}\Phi = 0 ,$$

in the sense of a distribution. From first equation it follows that

$$\Phi(\boldsymbol{x}, t, \xi) = e^{-im_0\xi}\phi(\boldsymbol{x}, t) .$$

Writing $\phi = (\omega, \chi, \psi)$ with ψ and χ taking values in $\mathbb{C}^{2j+1}$ and ω in $\mathbb{C}^{2j-1}$, the second equation gives

$$\begin{aligned} m_0\omega + i\boldsymbol{T}^{(j-1)j} \cdot \boldsymbol{\mathcal{D}}\psi &= 0 \\ im_0\chi + i\boldsymbol{H}^j \cdot \boldsymbol{\mathcal{D}}\psi &= 0 \\ i(\boldsymbol{T}^{(j-1)j})^* \cdot \boldsymbol{\mathcal{D}}\omega + i\boldsymbol{H}^j \cdot \boldsymbol{\mathcal{D}}\chi + 2j^2 i\mathcal{D}_4\psi &= 0. \end{aligned}$$

Solving these four equations in terms of ϕ, using (6.12) to compute the expression $H_a H_b + (T_a^{(j-1)j})^* T_b^{(j-1)j}$, and using the commutation relations

$$[\mathcal{D}_a, \mathcal{D}_b] = -iq\epsilon_{abc}\boldsymbol{B}_c\ , \qquad a, b, c = 1, 2, 3,$$

one has

$$\left(i\frac{\partial}{\partial t} - qV(\boldsymbol{x}, t) - \frac{1}{2m_0}\left(i\boldsymbol{\nabla} + q\boldsymbol{A}(\boldsymbol{x}, t)\right)^2 + \frac{q}{2jm_0}\boldsymbol{H}^j \cdot \boldsymbol{B}(\boldsymbol{x}, t)\right)\psi(\boldsymbol{x}, t) = 0. \tag{6.10}$$

The physical interpretation of this equation suggests that a quantum particle with spin j interacting with an electromagnetic field acquires an intrinsic magnetic momentum with gyromagnetic ratio $g = \frac{1}{j}$, as deduced by Hurley [22]. Although the form of the interaction between the particle and the electromagnetic field has its root in the relativistic framework, (6.10) follows from Galilean covariance requirements, as soon as one assumes the minimal coupling principle and the fact that the wave equation is of the Dirac type.

The case $j = 0$ can be treated with similar calculations. The result is that the wave equation for the scalar component is the usual wave equation for a spinless particle in the electromagnetic field.

6.3 The 2 + 1 Dimensional Case

From Sect. 5.3 of Chap. 5 we know that the elementary free particles in $2+1$ dimensions are described by irreducible unitary representations of the universal covering group $\overline{G}$ of the Galilei group of the form

$$U^{m,\lambda} = \mathrm{Ind}_{\overline{G}_{p_m}}^{\overline{G}}\left(p_m \times D^{m,\lambda}\right),$$

where $p_m = (\mathbf{0}, 0, m)$, $\overline{G}_{p_m} = \mathbb{R}^4 \times' (\{\mathbf{0}\} \times \mathbb{R}^2)$ and $D^{m,\lambda}$ is the (scalar) representation of $\overline{G}_{p_m}$ given by

$$(\boldsymbol{a}, b, c, \mathbf{0}, r, x) \mapsto e^{i(mc+\lambda x)}.$$

We show that, if $\lambda \neq 0$, there are no wave equations associated with $U^{m,\lambda}$. Indeed, from Condition (3) of Theorem 5, with $p_0 = p_m$ and $\tau = D^{m,\lambda}$, it follows that the representation L of H, restricted to the stability subgroup $\{\mathbf{0}\} \times \mathbb{R}^2$, has to contain the character

$$(\mathbf{0}, r, x) \mapsto e^{i\lambda x}$$

as a subrepresentation. Due to the following Lemma, this condition implies that $\lambda = 0$.

Lemma 13. *Let L be a finite dimensional representation of H, then for all $x \in \mathbb{R}$, $L(\mathbf{0}, 0, x)$ is a unipotent operator.*

Proof. Let Lie $(H) = \mathbb{R}^2 \oplus \mathbb{R} \oplus \mathbb{R}$ be the Lie algebra of H and (e_1, e_2, e_3, e_4) the canonical basis. Define $\widehat{L}$ as the differential of L, which is a Lie algebra representation of Lie (H). Let l be an eigenvalue of $\widehat{L}(e_4)$ and W the corresponding eigenspace. Since $[e_1, e_4] = [e_2, e_4] = 0$, it follows that $\widehat{L}(e_1)$ and $\widehat{L}(e_2)$ leave invariant W, but $[e_1, e_2] = e_4$, so that $\widehat{L}(e_4)$ restricted to W has null trace and, hence, $l = 0$. $\square$

For fixed $m \neq 0$, we are thus looking for wave equations associated with $U^{m,0}$. Since we are interested in coupling the particle with an external electromagnetic field, we choose the representation L of the group H in such a way that there are L-invariant Dirac type wave equations.

Let L be the representation of H acting on $V = \mathbb{C}^2$ as

$$L_\pm(\boldsymbol{v}, r, x) = \begin{pmatrix} e^{\pm ir} & (v_x \pm i v_y) \\ 0 & 1 \end{pmatrix}.$$

As a consequence of Theorem 5, the following result is obtained.

Corollary 6. *Given $m \neq 0$, let D_1 and D_2 be the following differential operators*

$$D_1 = (i\frac{\partial}{\partial \xi} - m) I_2 \,,$$

$$D_2 = i \begin{pmatrix} \frac{\partial}{\partial \xi} & (\frac{\partial}{\partial x} \pm i\frac{\partial}{\partial y}) \\ (\frac{\partial}{\partial x} \mp i\frac{\partial}{\partial y}) & 2(\frac{\partial}{\partial t} - \varepsilon_0) \end{pmatrix}.$$

Then (D_1, D_2) is an $L_\pm$-invariant wave equation of the Dirac type.

The proof of the above corollary is very similar to that of Proposition 11 and we omit it. In [10], it is also shown that the above wave equations are unique, if one requires the dimension of the space, where L acts, to be at most 2.

If the interaction with a given external electromagnetic field $(\boldsymbol{E}, B_z)$ is again introduced by the minimal coupling, the computations of Sect. 6.2.1 can be repeated to yield the equation

$$\left(i\frac{\partial}{\partial t} - qV(\boldsymbol{x}, t) - \frac{1}{2m}\left(i\boldsymbol{\nabla} + q\boldsymbol{A}(\boldsymbol{x}, t)\right)^2 \mp \frac{q}{2m} B_z(\boldsymbol{x}, t)\right) \phi(\boldsymbol{x}, t) = 0, \tag{6.11}$$

where ϕ is a smooth function from $\mathbb{R}^3$ to $\mathbb{C}$.

From a physical point of view, comparing (6.11) with the corresponding three dimensional equation (6.10), we see that the two dimensional elementary particles of mass m (recall that $\lambda = 0$) interact with the electromagnetic field as three dimensional spin-1/2 particles with spin down (L_+-invariant wave equations) or spin up (L_--invariant wave equations). Notice that, since the *spin* of a $2+1$ dimensional particle has no physical meaning (in particular for the representation $U^{m,0}$ it is zero), the gyromagnetic ratio can not be defined.

We stress that this result depends on the minimal assumption on the dimension of the vector space carrying the representation L and it will have to be checked experimentally. This *minimal hypothesis* in the choice of L holds also for the relativistic Dirac equation whose nonrelativistic limit gives rise to the known gyromagnetic ratio $\frac{1}{j}$.

6.4 Finite Dimensional Representations of the Euclidean Group

To close our treatise we consider the finite dimensional continuous representations of the covering group $E(3)^*$ of the Euclidean group $E(3)$ in three dimensions. The group $E(3)^*$ is given by the semidirect product of the vector group $\mathbb{R}^3$ and the Lie group $SU(2)$ acting on $\mathbb{R}^3$ in the usual way. It is evident that $E(3)^*$ coincides with the homogeneous factor H of the universal central extension of the Galilei group in 3+1 dimensions, as defined in Sect. 4.1 of Chap. 4. Finite dimensional representations are needed in constructing Galilei invariant wave equations. Hence, in the following, let $H = E(3)^*$.

Since H is simply-connected, its finite dimensional representations are in one to one correspondence with the representations of its Lie algebra. One has that $\mathrm{Lie}\,(H)$ is the semidirect sum of the Abelian Lie algebra $\mathbb{R}^3$ and the simple Lie algebra $su(2)$. Due to the fact that $\mathrm{Lie}\,(H)$ is neither compact nor semisimple, it is very difficult to obtain a complete classification of the representations of $\mathrm{Lie}\,(H)$. Nevertheless George and Lévy-Nahas [16], reduce the problem to the one of solving a non linear matrix equation, see relation (6.13) below. In the following we briefly describe their results.

Let $\boldsymbol{H} = (H_1, H_2, H_3)$ be the usual basis of $su(2)$ and denote the elements of $\mathrm{Lie}\,(H)$ as $(\boldsymbol{v}, \boldsymbol{l})$, with $\boldsymbol{v}, \boldsymbol{l} \in \mathbb{R}^3$, instead of $(\boldsymbol{v}, \boldsymbol{l}\cdot\boldsymbol{H})$.

For each $j \in \frac{1}{2}\mathbb{N}$, let d^j be the irreducible representation of $su(2)$ labeled by j and let

$$\boldsymbol{H}^j = \left(id^j(H_1), id^j(H_2), id^j(H_3)\right)$$

be the corresponding Hermitian generators.

For all $j, k \in \frac{1}{2}\mathbb{N}$, the dimension of the vector space

$$W^{1,k,j} := \mathrm{Hom}_{SU(2)}(\mathbb{R}^3; \mathcal{L}(\mathbb{C}^{2j+1}, \mathbb{C}^{2k+1}))$$

is one if $|k - j| \leq 1$, whereas it is zero in the other cases. Hence, let $T^{kj} \in W^{1,k,j}$ be normalized in such a way that

$$\begin{aligned} \boldsymbol{T}^{jj} &= \boldsymbol{H}^j \\ \boldsymbol{T}^{j(j-1)} &= (\boldsymbol{T}^{(j-1)j})^* \\ \boldsymbol{v}_1 \cdot \boldsymbol{H}^j \boldsymbol{v}_2 \cdot \boldsymbol{H}^j + \boldsymbol{v}_1 \cdot (\boldsymbol{T}^{j(j-1)})^* \boldsymbol{v}_2 \cdot \boldsymbol{T}^{(j-1)j} &= i\, j(\boldsymbol{v}_1 \wedge \boldsymbol{v}_2) \cdot \boldsymbol{H}^j + j^2 \boldsymbol{v}_1 \cdot \boldsymbol{v}_2, \end{aligned} \tag{6.12}$$

where $\boldsymbol{v} \cdot \boldsymbol{T}^{(j-1)j} := T^{(j-1)j}(\boldsymbol{v})$. The explicit form of the matrices $\boldsymbol{T}^{(j-1)j}$ can be found in [15]. The relation (6.12) is fundamental for the problem of finding a Dirac type wave equation.

Let ρ be a finite dimensional representation of Lie (H). One can prove, [16], that ρ can be written, in a suitable basis, in upper triangular block form of the type:

$$\rho(\boldsymbol{v}, \boldsymbol{l}) = \begin{pmatrix} \rho_1(\boldsymbol{l}) & \widetilde{f}_{12}(\boldsymbol{v}) & \dots & \widetilde{f}_{1n}(\boldsymbol{v}) \\ 0 & \rho_2(\boldsymbol{l}) & \dots & \widetilde{f}_{2n}(\boldsymbol{v}) \\ \vdots & \vdots & \ddots & \vdots \\ 0 & 0 & \dots & \rho_n(\boldsymbol{l}) \end{pmatrix} .$$

The blocks on the diagonal are given by

$$\rho_p(\boldsymbol{l}) = \oplus_j (I_{\widetilde{n}_{jp}} \otimes (\boldsymbol{l} \cdot \boldsymbol{H}^j)) ,$$

where $\widetilde{n}_{jp}$ is the multiplicity of the j-representation in the p-block. The off-diagonal blocks are of the form, with $p < q$,

$$\widetilde{f}_{pq}(\boldsymbol{v}) = (\widetilde{f}_{ij,pq}(\boldsymbol{v}))_{ij} = (M_{ij,pq} \otimes \boldsymbol{v} \cdot \boldsymbol{T}^{ij})_{ij} ,$$

where $M_{ij,pq}$ are $\widetilde{n}_{ip} \times \widetilde{n}_{jq}$ matrices. One has to determine these matrices in a such way that the following non linear equation holds:

$$\sum_k \gamma_{ijk} M_{ik} M_{kj} = 0 , \tag{6.13}$$

where $M_{ij} = (M_{ij,pq})_{pq}$ is a block matrix $(\sum_p \widetilde{n}_{ip} \times \sum_q \widetilde{n}_{jq})$ whose elements are the matrices $M_{ij,pq}$ ($M_{ij,pq} = 0$ if $p \geq q$), and γ_{ijk} are known coefficients that can be found in [16]. We notice that in the case $\widetilde{n}_{ip} = 1$ for all i, p, the matrices $M_{ij,pq}$ are reduced to scalars.

Applying this technique, one obtains the representation L^j introduced by relation (6.8), with $2j \in \mathbb{N}$, $j \geq \frac{1}{2}$.

A Appendix

A.1 Dictionary of Mathematical Notions

This dictionary gives the definitions and the basic properties of most of the mathematical concepts that are freely used in the book. No references are given since the material is standard. In this Appendix $\mathcal{H}$ is a complex separable Hilbert space (see the corresponding item below).

Absolute value of an operator. The *absolute value* $|A|$ of an operator $A \in \mathbf{B}$ is the unique positive operator $|A| \in \mathbf{B}$ such that $|A|^2 = A^*A$.

Adjoint of an operator. The *adjoint* of an operator $A \in \mathbf{B}$ is the unique operator $A^* \in \mathbf{B}$ such that $\langle A^*\varphi, \psi\rangle = \langle \varphi, A\psi\rangle$ for all $\varphi, \psi \in \mathcal{H}$. The map $\mathbf{B} \ni A \mapsto A^* \in \mathbf{B}$ is an antilinear map such that $(A^*)^* = A$, $\|A^*\| = \|A\|$ and $\|A^*A\| = \|A\|^2$, that is, $\mathbf{B}$ is a C^*-algebra.

Analytic function. Let M be an analytic manifold. A function $f : M \to \mathbb{R}$ is *analytic* at the point $p \in M$ if there is a chart (U, φ) of M such that $p \in U$ and $f \circ \varphi^{-1} : \varphi(U) \to \mathbb{R}$ is real analytic. The set of analytic functions on M at the point p is a real vector space and we denote it by $\mathcal{F}(p)$.

Antilinear operator. An additive map $A : \mathcal{H} \to \mathcal{H}$ for which $A(c\varphi) = \bar{c}A\varphi$ for all $\varphi \in \mathcal{H}$ and $c \in \mathbb{C}$ (where $\bar{c}$ denotes the complex conjugate of c) is an *antilinear operator*.

Antiunitary operator. A bounded antilinear operator U is *antiunitary* if $UU^* = U^*U = I$, that is, if $U^* = U^{-1}$. The antiunitary operators are the bijective antilinear functions $U : \mathcal{H} \to \mathcal{H}$ which reverse the inner product, $\langle U\varphi, U\psi\rangle = \langle \psi, \varphi\rangle$ for all $\varphi, \psi \in \mathcal{H}$. An antilinear U is antiunitary if and only if it is an isometry and a surjection. We let $\overline{\mathbf{U}}$ denote the set of all antiunitary operators on $\mathcal{H}$. The product of two antiunitary operators is unitary. The set $\overline{\mathbf{U}}$ can be equipped by the norm topology as well as by the strong and weak operator topologies. In particular, the strong and weak operator topologies coincide (cp. unitary operators).

Atoms of D. An *atom* of the set $\mathbf{D}$ is an element $P \in \mathbf{D}$ for which the condition $O \leq D \leq P$, $D \in \mathbf{D}$, implies that either $D = O$ or $D = P$. They are exactly the one-dimensional projections on $\mathcal{H}$. Any $D \in \mathbf{D}$ is the least upper bound of the atoms contained in it, $D = \vee_{P \leq D} P$. Since $\mathcal{H}$ is separable, any D can be expressed as the least upper bound of at most countably many atoms contained in D (cp. weak atom).

G. Cassinelli, E. De Vito, P.J. Lahti, and A. Levrero, *The Theory of Symmetry Actions in Quantum Mechanics*, Lect. Notes Phys. **654**, pp. 89–101
http://www.springerlink.com/

Borel measure. Let X be an lcsc space. The Borel σ-algebra $\mathcal{B}(X)$ of X is the σ-algebra of subsets of X generated by its open sets. A measure $\mu : \mathcal{B}(X) \to [0,\infty]$ that is finite on the compact sets is called a *Borel measure* on X. Any Borel measure on X is necessarily σ-finite, that is, X can be expressed as a countable union of disjoint sets $E_n \in \mathcal{B}(X)$ for which $\mu(E_n) < \infty$, and regular, that is, for any $E \in \mathcal{B}(X)$, $\mu(E) = \sup\{\mu(K) \,|\, K \subset E \text{ compact}\} = \inf\{\mu(O) \,|\, E \subset O \text{ open}\}$.

Bounded antilinear operator. An antilinear operator A is bounded if there is a constant $M \in [0,\infty)$ such that $\|A\varphi\| \leq M \|\varphi\|$ for all $\varphi \in \mathcal{H}$. The norm of a bounded antilinear operator A is given as $\|A\| := \sup\{\|A\varphi\| \mid \varphi \in \mathcal{H}, \|\varphi\| \leq 1\}$ $(< \infty)$. The adjoint A^* of an antilinear operator A is the unique antilinear operator for which $\langle \psi, A^*\varphi \rangle = \langle \varphi, A\psi \rangle$ for all $\varphi, \psi \in \mathcal{H}$.

Bounded operator. A linear operator $A : \mathcal{H} \to \mathcal{H}$ is a *bounded (linear) operator* if there is a constant $M \in [0,\infty)$ such that $\|A\varphi\| \leq M \|\varphi\|$ for all $\varphi \in \mathcal{H}$. The existence of such a constant is equivalent to the continuity of A. We let $\mathbf{B}$ denote the set of all linear bounded operators on $\mathcal{H}$. If $\mathcal{H}$ is to be emphasized, we denote it as $\mathbf{B}(\mathcal{H})$.

Character. A *character* of an lcsc group G is a continuous (group) homomorphism $G \to \mathbb{T}$. The set of characters of G is a group under pointwise multiplication. When G is an Abelian group the group of characters of G is denoted by $\hat{G}$ and is called the *dual group of* G. The group $\hat{G}$ is an lcsc group with respect to topology of the uniform convergence on compact sets.

Compact selfadjoint operator. An operator $A \in \mathbf{B}$ is *compact* if the closure of the set $\{A\varphi \,|\, \|\varphi\| \leq 1\}$ is compact. The spectral structure of compact selfadjoint operators is particularly simple. Indeed, if A is a compact selfadjoint operator, then A can be expressed as a norm convergent series (or as a finite sum) $A = \sum_n \lambda_n P_n$, where $\lambda_n \neq 0$ for each n, P_n is a nonzero projection (of finite rank) for each n, $P_n P_m = 0$ for $n \neq m$, and $\lambda_n \neq \lambda_m$ for $n \neq m$. Moreover, the set of numbers λ_n in the formula $A = \sum_n \lambda_n P_n$ is the set $\sigma(A) \setminus \{0\}$, where $\sigma(A)$ is the spectrum of A, and $\lim_{n\to\infty} \lambda_n = 0$, provided that the numbers λ_n are infinitely many.

Connected component. A *component* of a point x of an lcsc space X is the union of all connected subspaces of X containing x. If x is the identity element e of an lcsc group G and G_e is the *connected component* of e then, G_e is a closed normal subgroup of G.

Connected group. A topological group is *connected* if it is connected as a topological space, that is, if it is not the union of two nonempty open disjoint subsets. A topological subgroup is connected if it is connected with respect to the relative topology.

Cauchy-Schwarz inequality. The *Cauchy-Schwarz inequality* states that, for any two vectors $\varphi, \psi \in \mathcal{H}$,

$$|\langle \varphi, \psi \rangle| \leq \|\varphi\| \, \|\psi\| \, .$$

Equality of operators. Given two operators $A, B \in \mathbf{B}$, acting on a complex Hilbert space $\mathcal{H}$, $A = B$ if and only if $\langle \varphi, A\varphi \rangle = \langle \varphi, B\varphi \rangle$ for all $\varphi \in \mathcal{H}$.

Exponential map. Let G be a Lie group. The exponential map exp is the unique analytic map from Lie (G) to G such that

1. for all $X \in \text{Lie}\,(G)$, the map

$$\mathbb{R} \ni t \to \exp(tX)$$

is a group homomorphism of the real additive line into G;
2. for all $X \in \text{Lie}\,(G)$ and $f \in \mathcal{F}(G)$,

$$\frac{d\, f(\exp(tX)}{dt}_{|t=0} = X(f).$$

In particular, if G is a matrix group, so that $X \in \text{Lie}\,(X)$ is a matrix, then

$$\exp(X) = \sum_{n=0}^{\infty} \frac{1}{n!} X^n.$$

Fréchet-Riesz theorem. The Fréchet-Riesz theorem assures that for each bounded linear functional $f : \mathcal{H} \to \mathbb{C}$ there is a unique $\psi \in \mathcal{H}$ such that $f(\varphi) = \langle \psi, \varphi \rangle$ for all $\varphi \in \mathcal{H}$.

***G*-space.** Let G be an lcsc group. An lcsc space X is called an (lcsc) G-space if G acts on X by means of a continuous map, called *action*

$$G \times X \ni (g, x) \mapsto g[x] \in X \,,$$

such that

$$\begin{aligned} e[x] &= x \quad \text{for all } x \in X \,, \\ (g_1 g_2)[x] &= g_1[g_2[x]] \quad \text{for all } g_1, g_2 \in G,\, x \in X. \end{aligned}$$

In particular, for each $g \in G$ the mapping $X \ni x \mapsto g[x] \in X$ is a homeomorphism.

Let $x \in X$. The set $G_x = \{g \in G \,|\, g[x] = x\}$ is a closed subgroup of G, the *stability subgroup* at x, and the set $G[x] = \{g[x] \in X \,|\, g \in G\}$ is the *orbit* of the point x.

Haar measure. Let G be an lcsc group. A left [or right] Haar measure on G is a Borel measure which is invariant with respect to the left [or right] action of G on itself given by

$$g[h] := gh \quad \text{for all } g, h \in G \quad (\text{ or } [h]g := hg \;) \,.$$

A theorem of André Weil assures that left and right Haar measures exist and are unique up to a positive multiplicative constant. If the left Haar measure is also right invariant, the group is called unimodular. The Galilei group and its universal extension both in $3+1$ and $2+1$ dimensions are unimodular.

Hilbert basis. A *(Hilbert) basis* $(\xi_n)_{n\in\mathbb{I}}$ is a collection of mutually orthogonal vectors in $\mathcal{H}$, $\langle \xi_n, \xi_m \rangle = \delta_{n,m}$ for all $n, m \in \mathbb{I}$, such that their closed linear span $\overline{\text{lin}\,\{\xi_n\}} = \mathcal{H}$. Moreover, for all $\varphi \in \mathcal{H}$, it holds that

$$\begin{aligned} \varphi &= \sum_{n\in\mathbb{I}} \langle \xi_n, \varphi \rangle \xi_n \quad \text{Fourier series} \\ \|\varphi\|^2 &= \sum_{n\in\mathbb{I}} |\langle \xi_n, \varphi \rangle|^2 \quad \text{Parseval formula.} \end{aligned}$$

The cardinality of the index set $\mathbb{I}$ is the *dimension* of $\mathcal{H}$. Since $\mathcal{H}$ is separable, $\mathbb{I}$ is either a finite set or countable.

Hilbert space. A complex *Hilbert space* $\mathcal{H}$ is a vector space $\mathcal{H}$ over $\mathbb{C}$ with an inner product $\langle \cdot, \cdot \rangle$, linear in the second argument, such that $\mathcal{H}$ is complete with respect to the norm

$$\|\varphi\| := \sqrt{\langle \varphi, \varphi \rangle}\,, \quad \varphi \in \mathcal{H}.$$

A Hilbert space $\mathcal{H}$ is *separable* if it has a countable dense subset. In this book by *Hilbert space* we mean a complex separable Hilbert space.

Invariant measure. Let X be a G-space. If μ is a Borel measure on X then the image measure $g(\mu)$ of μ under the mapping $x \mapsto g[x]$ is defined as

$$g(\mu)(E) = \mu(\{x : g[x] \in E\})$$

for each $E \in \mathcal{B}(X)$. In an equivalent way, $g(\mu)$ can be defined by the equality

$$\int_X f(x) dg(\mu)(x) = \int_X f(g[x]) d\mu(x)$$

for all $f \in C_c(X)$, where $C_c(X)$ is the set of continuous functions with compact support.

The measure μ is called *invariant* if $\mu = g(\mu)$ for all $g \in G$. The invariance of μ is equivalent to any of the following conditions:

a) $\mu(E) = g(\mu)(E)$ for all $E \in \mathcal{B}(X)$;
b) $\mu(K) = g(\mu)(K)$ for all $K \subset X$ compact;
c) $\int_X f(g[x]) d\mu(x) = \int_X f(x) d\mu(x)$ for all $f \in C_c(X)$.

If G is a unimodular and X is a transitive G-space such that the stability subgroup is also unimodular, then X admits an invariant measure, unique up to a positive multiplicative constant.

Irreducible unitary representation. A unitary representation U of a group G acting on $\mathcal{H}$ is *irreducible* if the null space and the whole space are the only invariant closed subspaces. The Schur lemma assures that U is irreducible if and only if the only operators in $\mathbf{B}$ commuting with all U_g, $g \in G$, are the ones proportional to the identity.

Lcsc space and group. A topological Hausdorff space is called *locally compact second countable* (lcsc) if each point has a compact neighborhood and it satisfies the second axiom of countability. An *lcsc* group is a topological Hausdorff group with an lcsc topology.

Lie algebra. A (real) *Lie algebra* $\mathfrak{g}$ is a vector space over $\mathbb{R}$ endowed with an antisymmetric bilinear mapping (Lie bracket) $\mathfrak{g} \times \mathfrak{g} \ni (X, Y) \mapsto [X, Y] \in \mathfrak{g}$ satisfying the Jacobi identity,

$$[[X, Y], Z] + [[Y, Z], X] + [[Z, X], Y] = 0.$$

Lie algebra homomorphism. Let $\mathfrak{g}_1$ and $\mathfrak{g}_2$ be two Lie algebras. A (*Lie algebra*) *homomorphism* $f : \mathfrak{g}_1 \to \mathfrak{g}_2$ is a linear function which preserves the Lie bracket, that is,

$$[f(X), f(Y)]_2 = f\left([X, Y]_1\right).$$

An isomorphism of Lie algebras is a bijective homomorphism.

Lie algebra of a Lie group. Let G be a Lie group G. A vector field $X \in D_1(G)$ is *left invariant* if the following condition holds. For all $g \in G$ and $f \in \mathcal{F}(G)$,

$$X(f^g) = X(f)^g,$$

where f^g is the function defined by $f^g(h) = f(g^{-1}h)$, for all $h \in G$.

Left invariant vector fields form a subalgebra $\mathrm{Lie}\,(G)$ of the Lie algebra $D_1(G)$ and $\mathrm{Lie}\,(G)$ is called the *Lie algebra* of G. A standard result of the theory of Lie groups assures that $\mathrm{Lie}\,(G)$ is isomorphic to the tangent space $T_e(G)$ of G at the identity e by means of

$$\mathrm{Lie}\,(G) \ni X \mapsto X_e \in T_e(G).$$

In particular, $\mathrm{Lie}\,(G)$ is a finite dimensional vector space.

Lie Group. An lcsc group G is a (real) *Lie group* if there is a (real) analytic structure on the set G, compatible with its topology, which converts it into a (real analytic) manifold and for which the group operations $(g, h) \mapsto gh$ and $g \mapsto g^{-1}$ are analytic. If G is a Lie group, then G, as a topological group, is an lcsc group.

Lie group homomorphism. Let G_1 and G_2 be two Lie groups. A *(Lie group) homomorphism* $\pi : G_1 \to G_2$ is a group homomorphism which is also an analytic mapping of the manifold underlying G_1 into the manifold underlying G_2. A (Lie group) isomorphism is a bijective Lie group homomorphism such that the inverse is also a Lie group homomorphism.

A Lie group homomorphism $\pi : G_1 \to G_2$ defines a Lie algebra homomorphism $\dot{\pi} : \mathrm{Lie}\,(G_1) \to \mathrm{Lie}\,(G_2)$ in the following way. Given $X \in \mathrm{Lie}\,(G_1)$, $\dot{\pi}(X)$ is the left invariant vector field on G_2

$$\mathcal{F}(G_2) \ni \mapsto X(f \circ \pi).$$

The following converse result holds.

Theorem 6. *Let G_1 and G_2 be connected Lie groups and f : Lie$(G_1) \to$ Lie(G_2) a Lie algebra homomorphism. If G_1 is simply connected, then there exists one and only one Lie group homomorphism $\pi : G_1 \to G_2$ such that $\dot{\pi} = f$.*

Lie groups and Lie algebras: main theorems. There are two results due to Sophus Lie about the structure of Lie groups.

Theorem 7. *Let $\mathfrak{g}$ be a Lie algebra. Then there is a connected, simply connected Lie group whose Lie algebra is isomorphic to $\mathfrak{g}$.*

Theorem 8. *Let G_1 and G_2 be Lie groups and Lie(G_1) and Lie(G_2) the corresponding Lie algebras. Then Lie(G_1) and Lie(G_2) are isomorphic if and only if G_1 and G_2 are locally analytically isomorphic, that is, if there exist two open neighborhoods U_1 and U_2 of the identities in G_1 and G_2 and an analytic diffeomorphism f of U_1 onto U_2 such that for any $g, h \in U_1$, we have that $gh \in U_1$ if and only if $f(g)f(h) \in U_2$ and, if this is the case, $f(gh) = f(g)f(h)$.*

Lie subgroup. Let G be a Lie group of dimension n. An (algebraic) subgroup H of G is called a *Lie subgroup* (of dimension $m < n$) if the following condition holds: for all $h_0 \in H$, there is a chart (U, φ) of G such that $h_0 \in U$, $\varphi(h_0) = 0$ and $\varphi(U \cap H)$ is the intersection of the open set $\varphi(U) \subset \mathbb{R}^n$ and an m-dimensional vectorial subspace of $\mathbb{R}^n$. In this case, on H there is a unique real analytic structure compatible with the relative topology such that H is a Lie group and the canonical immersion $i : H \to G$ is analytic.

A Lie subgroup H is always a closed subgroup of G and its Lie algebra Lie(H) is a Lie subalgebra of Lie(G). Conversely, any closed subgroup of G is a Lie subgroup. We notice that in the literature there are different not equivalent definitions of Lie subgroups. Our definition is strong enough to assure that a Lie subgroup is always closed in G, compare with the definition of a *regularly embedded Lie subgroup* of [38].

Manifold. Let M be an lcsc space. A *chart* of M is a pair (U, φ), where U is an open set of M and φ is a homeomorphism of U onto an open subset of $\mathbb{R}^n$ for some n. The number n is the dimension of the chart. Different charts on M have the same dimension. A real *analytic structure* on M is a set $\{(U_i, \varphi_i) \mid i \in \mathbb{I}\}$, $\mathbb{I}$ an index set, where for each $i \in \mathbb{I}$, the pair (U_i, φ_i) is a chart of dimension n on M such that $\cup_i U_i = M$ and for each $i, j \in \mathbb{I}$ the map $\varphi_j \circ \varphi_i^{-1} : \varphi_i(U_i \cap U_j) \to \varphi_j(U_i \cap U_j)$ is a real analytic function. We say that M is *a (real analytic) manifold* of dimension n if a real analytic structure is defined on M.

Measurable function. Let X and Y be two lcsc spaces. A map $f : X \to Y$ is called measurable if, for all $E \in \mathcal{B}(Y)$, $f^{-1}(E) \in \mathcal{B}(X)$.

Norm of an operator. Let $A \in \mathbf{B}$ be an operator. The norm of A is defined as $\|A\| := \sup\{\|A\varphi\| \mid \varphi \in \mathcal{H}, \|\varphi\| \leq 1\}$ and it satisfies $\|AB\| \leq \|A\| \, \|B\|$, for all $A, B \in \mathbf{B}$, that is, $\mathbf{B}$ is a Banach algebra.

One-dimensional projection. A projection P is a *one-dimensional projection* if it is a projection on a one-dimensional subspace of $\mathcal{H}$. If $\varphi \in \mathcal{H}, \varphi \neq 0$, then $P = P[\varphi]$, where $P[\varphi]\psi := \frac{\langle \varphi, \psi \rangle}{\langle \varphi, \varphi \rangle}\varphi$, for all $\psi \in \mathcal{H}$. Clearly, $P[\varphi] = P[\psi]$ if and only if $\varphi = c\psi$ for some $c \in \mathbb{C}$, $c \neq 0$. We let $\mathbf{P}$ denote the set of all one-dimensional projections.

Operator order. For any $A, B \in \mathbf{B}$ we write $A \leq B$, and say that A is contained in B, if $B - A$ is positive. The relation $\leq$ is an *order* on $\mathbf{B}$, and it makes $\mathbf{B}$ a partially ordered vector space. We recall that $\mathbf{B}$ is an antilattice, that is, any two elements $A, B \in \mathbf{B}$ have the greatest lower bound $A \wedge B$ in $\mathbf{B}$ if and only if A and B are comparable, that is, either $A \leq B$ or $B \leq A$.

Orthogonal vectors. Two vectors $\varphi, \psi \in \mathcal{H}$ are *orthogonal*, $\varphi \perp \psi$, if $\langle \varphi, \psi \rangle = 0$, and a set $\mathcal{K} \subset \mathcal{H}$ is *orthonormal* if the vectors $\varphi \in \mathcal{K}$ are mutually orthogonal unit vectors.

Polarization identity. *The polarization identity* $\langle \varphi, \psi \rangle = \frac{1}{4}\sum_{n=0}^{3} i^n \|\psi + i^n\varphi\|$, $\varphi, \psi \in \mathcal{H}$, connects the inner product and the norm of a Hilbert space $\mathcal{H}$.

Positive operator. An element $A \in \mathbf{B}$ is *positive*, $A \geq O$, if $\langle \varphi, A\varphi \rangle \geq 0$ for all $\varphi \in \mathcal{H}$. Positive operators are selfadjoint. We let $\mathbf{B}^+$, or, equivalently, $\mathbf{B}_r^+$, denote the set of all positive operators on $\mathcal{H}$.

Projection operator and the projection lattice. An operator $D \in \mathbf{B}$ is a *projection* if $D = D^2 = D^*$. We let $\mathbf{D}$ denote the set of all projections on $\mathcal{H}$. When the order on $\mathbf{B}$ is restricted on $\mathbf{D}$, $\mathbf{D}$ gains the structure of a complete lattice with the zero operator O and the unit operator I as the order bounds, $O \leq D \leq I$ for all $D \in \mathbf{D}$. The map $D \mapsto D^\perp := I - D$ is an orthocomplementation and it turns $\mathbf{D}$ into a complete orthocomplemented orthomodular lattice.

Projections and closed subspaces. The set $\mathbf{D}$ of projections on $\mathcal{H}$ stands in one to one onto correspondence with the set $\mathbf{M}$ of closed subspaces of $\mathcal{H}$. If $D \in \mathbf{D}$, then its range $D(\mathcal{H}) := \{D\varphi \,|\, \varphi \in \mathcal{H}\}$ is a closed subspace. On the other hand, if $M \subseteq \mathcal{H}$ is a closed subspace, then $\mathcal{H} = M \oplus M^\perp$, where $M^\perp := \{\psi \in \mathcal{H} \,|\, \psi \perp \xi \text{ for all } \xi \in M\}$. Hence, each $\varphi \in \mathcal{H}$ can uniquely be expressed as $\varphi = \varphi_M + \varphi_{M^\perp}$, with $\varphi_M \in M$, $\varphi_{M^\perp} \in M^\perp$. Then $D_M : \varphi \mapsto \varphi_M$ is a projection, with $D_M(\mathcal{H}) = M$. The correspondence $D \mapsto D(\mathcal{H})$, or, inversely, $M \mapsto D_M$, is a bijection, and it preserves both the order ($D_1 \leq D_2 \Leftrightarrow D_1(\mathcal{H}) \subseteq D_2(\mathcal{H})$) and the orthocomplementation ($D(\mathcal{H})^\perp = D^\perp(\mathcal{H})$).

Quotient space. Let G be an lcsc group and H a closed subgroup. The quotient space G/H is the set of equivalence classes of G with respect to the following relation:

$$g_1 \sim g_2 \iff \text{there is an } h \in H \text{ such that } g_2 = g_1 h\ .$$

The quotient space G/H, endowed with the quotient topology, is a transitive G-space with respect to the action

$$g_1[\dot{g}] = g_1 g, \quad g \in G, \dot{g} \in G/H\ ,$$

where $\dot{g}$ denotes the equivalence class of g. In particular the stability subgroup at $\dot{e}$ is H.

Section. Let G be an lcsc group and X a transitive G-space. Let $x_o \in X$, a section is a map $c : X \to G$ such that $c(x_o) = e$ and $c(x)[x_o] = x$ for all $x \in G[x_o]$. A result of George Mackey assures that there always exists a measurable section.

Selfadjoint operator. An operator $A \in \mathbf{B}$ is called *selfadjoint* if $A^* = A$ or, equivalently, if $\langle \varphi, A\varphi \rangle \in \mathbb{R}$ for all $\varphi \in \mathcal{H}$. We let $\mathbf{B}_r$ denote the set of all selfadjoint operators on $\mathcal{H}$. If $A \in \mathbf{B}_r$ there is a *spectral measure* $\Pi^A : \mathcal{B}(\mathbb{R}) \to \mathbf{B}$ such that $\Pi^A([-\|A\|, \|A\|]) = I$ and , for any $\varphi \in \mathcal{H}$,

$$\langle \varphi, A\varphi \rangle = \int_{\mathbb{R}} x \, d\Pi^A_{\varphi,\varphi}(x).$$

Semidirect product. Let A, H be two Lie groups and assume that H acts on A in such a way that

1. for all $h \in H$, the map $a \mapsto h[a]$ is a group homomorphism;
2. the map $(a, h) \mapsto h[a]$ from $A \times H$ to A is analytic.

The product manifold $G = A \times H$ becomes a Lie group with respect to the composition law

$$(a, h)(a', h) := (ah[a']), hh') \quad (a, h), (a', h') \in A \times H, \tag{A.1}$$

The group G is called the *semidirect product* of A and H and it is denoted by $A \times' H$. The groups A and H are canonically identified with closed subgroups of G is such a way that

$$A \cap H = \{e\} , \tag{A.2}$$
$$AH = G , \tag{A.3}$$
$$hAh^{-1} \subset A . \tag{A.4}$$

(Equivalently, (A.4) says that A is a *normal* subgroup of G). Conversely, given a Lie group G and two closed subgroups A and H such that (A.2)–(A.4) hold, then G is (isomorphic to) the semidirect product of A and H with respect to the canonical action of H on A given by

$$h[a] = hah^{-1},$$

which is the *inner action.*

Simply connected group. Let X be a manifold. A *path* is a continuous map $p : [0, 1] \to X$. The space X is said to be *simply connected* if the following condition holds. For all paths p and q such that $p(0) = q(0) = x$ and $p(1) = q(1) = y$ there is a continuous map $\Xi : [0, 1] \times [0, 1] \to X$ such that

$$\begin{aligned} \Xi(0,t) &= p(t) \quad t \in [0,1] \\ \Xi(1,t) &= q(t) \quad t \in [0,1] \\ \Xi(s,0) &= x \quad\ \ s \in [0,1] \\ \Xi(s,1) &= y \quad\ \ s \in [0,1]. \end{aligned}$$

A Lie group is *simply connected* if it is simply connected as a manifold.

Spectral measure. A (real) spectral measure (or projection valued measure) is a map Π from the Borel σ-algebra $\mathcal{B}(\mathbb{R})$ of the real line $\mathbb{R}$ into the set $\mathbf{B}$ of bounded operators on $\mathcal{H}$ such that

$$\begin{aligned} &\Pi(X) \in \mathbf{D} \quad \text{for all } \mathbf{X} \in \mathcal{B}(\mathbb{R}), \\ &\Pi(\mathbb{R}) = I, \\ &\Pi(\cup_i X_i) = \sum_i \Pi(X_i), \end{aligned}$$

for all sequences $(X_i)_{i \in I}$ of disjoint sets in $\mathcal{B}(\mathbb{R})$ (with the series converging in the strong, or equivalently, in the weak operator topology). Equivalently, a map $\Pi : \mathcal{B}(\mathbb{R}) \to \mathbf{D}$ is a spectral measure if for each unit vector $\varphi \in \mathcal{H}$, the map $X \mapsto \langle \varphi, \Pi(X)\varphi \rangle =: \Pi_{\varphi,\varphi}$ is a probability measure.

Strong operator topology. The *strong operator topology* on $\mathbf{B}$ is the weakest topology with respect to which all the functions $\mathbf{B} \ni A \mapsto A\varphi \in \mathcal{H}$, $\varphi \in \mathcal{H}$, are continuous. A net $(A_i)_{i \in \mathcal{I}}$ of bounded operators converges to an operator $A \in \mathbf{B}$ strongly if $\lim A_i\varphi = A\varphi$ for all $\varphi \in \mathcal{H}$.

Tangent space. Let M be a real manifold of dimension n and $p \in M$. A *tangent vector* at p is a linear map $L : \mathcal{F}(p) \to \mathbb{R}$ which is also a derivation, that is, $L(fg) = L(f)\, g(p) + f(p)\, L(g)$ for all $f, g \in (p)$.

Topological group. A set G is a *topological group* if it is an abstract group and a topological space with the Hausdorff topology such that the group operations $(g,h) \mapsto gh$ and $g \mapsto g^{-1}$ are continuous.

Topology on U. The set $\mathbf{U}$ of unitary operators is a closed subset of $\mathbf{B}$ in the weak operator topology. However, when restricted on $\mathbf{U}$ the weak and strong operator topologies coincide.

Torus $\mathbb{T}$. Let $\mathbb{T} = \{z \in \mathbb{C} \,|\, |z| = 1\}$ denote the set of complex numbers of modulus one. It is a multiplicative Lie group. We call it *the phase group* or *the torus.*

Trace class operators. An operator $T \in \mathbf{B}$ is of *trace class* if there is a basis $\mathcal{K}$ of $\mathcal{H}$ such that $\sum_{\xi \in \mathcal{K}} \langle \xi, |T|\xi \rangle < \infty$, where $|T|$ is the absolute value of T. We let $\mathbf{B}_1$ denote the set of all trace class operators on $\mathcal{H}$. If $T \in \mathbf{B}_1$, the series $\sum_{\varphi \in \mathcal{K}} \langle \varphi, T\varphi \rangle$ is absolutely convergent and the number $\mathrm{tr}[T] := \sum_{\varphi \in \mathcal{K}} \langle \varphi, T\varphi \rangle$ is the *trace* of $T \in \mathbf{B}_1$ (the definition of trace class operator and trace is independent of the choice of the basis $\mathcal{K}$). The trace is a linear functional on $\mathbf{B}_1$ and $\mathrm{tr}[AT] = \mathrm{tr}[TA]$ for any $A \in \mathbf{B}, T \in \mathbf{B}_1$ (this means that $\mathbf{B}_1$ is a *-ideal of $\mathbf{B}$).

Trace norm. The function $T \mapsto \|T\|_1 := \mathrm{tr}\big[|T|\big]$ is a norm, *the trace norm* on $\mathbf{B}_1$, and it turns $\mathbf{B}_1$ into a Banach space. For any $A \in \mathbf{B}, T \in \mathbf{B}_1$,

$|\mathrm{tr}[AT]| \leq \|A\| \, \|T\|_1$ and $\|T\| \leq \|T\|_1$. The dual space $\mathbf{B}_1^*$ of $(\mathbf{B}_1, \|\cdot\|_1)$ is isometrically isomorphic with the Banach space $(\mathbf{B}, \|\cdot\|)$, the duality being given by the function $\mathbf{B} \ni A \mapsto f_A \in \mathbf{B}_1^*$, where the functional f_A is defined by the formula $f_A(T) := \mathrm{tr}[AT]$ for all $T \in \mathbf{B}_1$.

Transitive G-space. Let G be an lcsc group and X be a G-space. If for each $x, y \in X$ there is a $g \in G$ such that $g[x] = y$, we say that X is a *transitive* G-space. If $x \in X$, then the orbit $G[x] = G$ and the map $G/G_x \ni \dot{g} \mapsto g[x] \in X$ is a homeomorphism of G-space, where G_x is the stability subgroup of G and G/G_x is the quotient space.

Vector field. Let M be a manifold. A (real analytic) *vector field* on M is a map $p \mapsto X_p$ that assigns to each point $p \in M$ a tangent vector X_p at the point p such that, for all $f \in \mathcal{F}(p)$, the function $M \ni p \mapsto X_p(f) \in \mathbb{R}$ is analytic.

Given a vector field X on M, the map $\mathcal{X} : \mathcal{F}(M) \to \mathcal{F}(M)$ given by

$$\mathcal{X}(f)(p) = X_p(f), \quad p \in M, \ f \in \mathcal{F}$$

defines a derivation, that is, a linear map on $\mathcal{F}(M)$ such that

$$\mathcal{X}(fg) = \mathcal{X}(f)g + f\mathcal{X}(g), \qquad f, g \in \mathcal{F}(M).$$

Conversely, any derivation on $\mathcal{F}(M)$ is of the above form and the correspondence between vector fields and derivation is one to one. The set of all vector fields (or derivation) is a real vector space denoted by $D_1(M)$ that becomes a Lie algebra with respect to the following Lie brackets: if $X, Y \in D_1(M)$, $[X, Y]$ is the vector field given by

$$f \mapsto [X, Y](f) := X(Y(f)) - Y(X(f)).$$

von Neumann theorem. The following result is due to John von Neumann.

Lemma 14. *Let G be an lcsc group and M a second countable topological group. Let $m : G \to M$ be a group homomorphism. Then m is continuous if and only if it is measurable.*

Unit vector. We say that $\varphi \in \mathcal{H}$ is a *unit vector* if $\|\varphi\| = 1$.

Unitary operator. An operator $U \in \mathbf{B}$ is *unitary* if one of the following equivalent conditions is satisfied

1. $UU^* = U^*U = I$;
2. U is bijective and $\langle U\varphi, U\psi \rangle = \langle \varphi, \psi \rangle$ for all $\varphi, \psi \in \mathcal{H}$, that is $U^{-1} = U^*$;
3. U is surjective and $\|U\varphi\| = \|\varphi\|$ for all $\varphi \in \mathcal{H}$.

We let $\mathbf{U}$ denote the set of all unitary operators on $\mathcal{H}$. See Sect. A.2 of Appendix A.1 for further details.

Unitarily equivalent representations. Unitary representations U and U' of G in Hilbert spaces $\mathcal{H}$ and $\mathcal{H}'$, respectively, are *unitarily equivalent* if there is a (linear) isometric isomorphism $V : \mathcal{H} \to \mathcal{H}'$ which intertwines the representations, that is, $VU_g = U'_g V$ for all $g \in G$.

Unitary representation. Let G be an lcsc group. A *unitary representation* of G in $\mathcal{H}$ is a map $G \ni g \mapsto U_g \in \mathbf{U}$ such that

1. $U_e = I$;
2. $U_{g_1 g_2} = U_{g_1} U_{g_2}$ for all $g_1, g_2 \in G$;
3. the map $g \mapsto U_g$ is continuous from G into $\mathbf{U}$ endowed with the strong (or, equivalently, weak) topology.

Lemma 14 and Proposition 12 of A.2 implies that $g \mapsto U_g$ is continuous if and only if, for all $\varphi, \psi \in \mathcal{H}$, the function $G \ni g \mapsto \langle \varphi, U_g \psi \rangle \in \mathbb{C}$ is measurable.

Universal covering group. Let G be a connected Lie group. There is a unique (up to an isomorphism) simply connected Lie group G^* and a (Lie group) surjective homomorphism $\delta : G^* \to G$ such that the kernel of δ is a discrete central closed subgroup of G^*. The group G^* is called *universal covering group* and δ the *covering homomorphism.*

Upper and lower bounds of operators. Let $\mathbf{C} \subset \mathbf{B}_r$. We say that $\mathbf{C}$ is bounded from above if it has an *upper bound*, that is, a $B \in \mathbf{B}$ such that $C \leq B$ for all $C \in \mathbf{C}$. If B_0 is an upper bound of $\mathbf{C}$ and $B_0 \leq B$ whenever B is an upper bound of $\mathbf{C}$, then B_0 is *the least upper bound*, and we denote $B_0 = \sup \mathbf{C}$, or $B_0 = \vee \mathbf{C}$. Similarly, one defines a *lower bound* and *the greatest lower bound* $\inf \mathbf{C}$, or $\wedge \mathbf{C}$. Let $(A_i)_{i \in \mathcal{I}} \subset \mathbf{B}_r$ be an increasing net, that is, $A_i \geq A_j$, when $i \geq j$. If the set $\{A_i \,|\, i \in \mathcal{I}\}$ is bounded from above, then it has the least upper bound A. Moreover, the net $(A_i)_{i \in \mathcal{I}}$ converges to A both weakly and strongly. A similar statement holds for decreasing nets that are bounded from below.

Weak atom. An element λP, $0 \leq \lambda \leq 1, P \in \mathbf{P}$, is called a *weak atom* of the set of unit bounded positive operators $O \leq E \leq I$ and any such operator E can be expressed as the join of the weak atoms contained in it, that is, $E = \vee_{\lambda P \leq E} \lambda P$ (cp. atoms of $\mathbf{D}$).

Weak operator topology. The *weak operator topology* is the weakest topology on $\mathbf{B}$ for which all the functions $\mathbf{B} \ni A \mapsto \langle \varphi, A\psi \rangle \in \mathbb{C}$, $\varphi, \psi \in \mathcal{H}$, are continuous. A net $(A_i)_{i \in \mathcal{I}}$ of bounded operators converges to an operator $A \in \mathbf{B}$ weakly if $\lim \langle \varphi, A_i \psi \rangle = \langle \varphi, A\psi \rangle$ for all $\varphi, \psi \in \mathcal{H}$.

A.2 The Group of Automorphisms of a Hilbert Space

In this appendix we briefly recall the mathematical properties of the set $\mathrm{Aut}\,(\mathcal{H})$ of automorphisms of a Hilbert space $\mathcal{H}$. We recall that an automorphism U of $\mathcal{H}$ is either a unitary operator or an antiunitary one, that is, $\mathrm{Aut}\,(\mathcal{H}) = \mathbf{U} \cup \overline{\mathbf{U}}$.

The main properties are stated by the following proposition, the proof of which is the same as the one of Lemma 5.34 and Lemma 5.4 of [35].

Proposition 12. *The set* Aut $(\mathcal{H})$ *is a group with respect to the usual composition between operators and it becomes a second countable metrisable topological group with respect to the strong operator topology. In particular,* $\mathbf{U}$ *is the connected component of the identity of* Aut $(\mathcal{H})$. *Finally, for a Borel space* X, *a function* $f : X \to \mathbf{U} \cup \overline{\mathbf{U}}$ *is measurable if and only if for all* $\varphi, \psi \in \mathcal{H}$ *the map* $X \ni x \mapsto \langle \varphi, f(x)\psi \rangle \in \mathbb{C}$ *is a measurable function.*

We define $\mathbf{T} := \{zI \mid z \in \mathbb{T}\}$. Clearly $\mathbf{T}$ is a closed central subgroup of Aut $(\mathcal{H})$ and it can be identified with the phase group $\mathbb{T}$.

Let Σ be the quotient group Aut $(\mathcal{H})/\mathbf{T}$. Its elements are the equivalence classes

$$[U] := \{U' \in \text{Aut}\,(\mathcal{H}) \mid U' = zU \text{ for some } z \in \mathbb{T}\}$$

and we let $\pi : \text{Aut}\,(\mathcal{H}) \to \Sigma, U \mapsto \pi(U) := [U]$ be the canonical projection.

We endow Σ with the quotient topology (we recall that $\Xi \subset \Sigma$ is open if and only if $\pi^{-1}(\Xi)$ is open in Aut $(\mathcal{H})$). The following corollary summarizes the basic properties of Σ and its proof is an easy consequence of the above proposition.

Corollary 7. *The group* Σ *is a second countable metrisable topological group and its connected component* Σ_0 *of the identity is* $\mathbf{U}/\mathbf{T}$. *In particular,* π *is a continuous open group homomorphism.*

Finally, we recall that a function $s : \Sigma_0 \to \mathbf{U}$ is a section for the canonical projection $\pi : \mathbf{U} \to \Sigma_0$ if $\pi \circ s = [I]$. If s is also a measurable function, it is called a measurable section.

The following result will be frequently used in the sequel, see Theorem 7.4 of [35]:

Proposition 13. *There is a measurable section* $s : \Sigma_0 \to \mathbf{U}$ *for the canonical projection* π *such that* s *is continuous in a neighborhood of the identity and* $s([I]) = I$.

A.3 Induced Representation

Here we briefly recall the definition of induced representation for semidirect products with a normal Abelian factor and its main properties (we refer to [35] for the proof).

Let G be a Lie group such that G is semidirect product of A and H where the normal factor A is Abelian. The dual group $\hat{A}$ of A has a natural structure of a manifold that converts it into a Lie group.

The action of G on A, $a \mapsto g[a] = gag^{-1}$, induces an action of G on $\hat{A}$, $x \mapsto g[x]$, which is defined through the following formula:

$$g[x](a) := x(g^{-1}[a]), \;\; a \in A, x \in \hat{A}, g \in G. \tag{A.5}$$

This action splits $\hat{A}$ into the orbits $G[x] := \{g[x] \mid g \in G\}$ of its points $x \in \hat{A}$.

To simplify the exposition, we assume that each orbit of $\hat{A}$ is locally closed (that is, the semidirect product is regular) and there is a G-invariant measure on each orbit.

Given $x_0 \in \hat{A}$, let

$$G_{x_o} = \{g \in G \,|\, g[x_o] = x_o\}$$

denote the *stability subgroups* at x_0 and

$$S_{x_o} = G_{x_o} \cap H \ ,$$

so that $G_{x_o} = A \times' S_{x_o}$. Let μ be a G-invariant measure on the orbit $G[x_0]$.

Given a unitary representation D of S_{x_o} acting on a Hilbert space $\mathcal{K}$, define the unitary representation $x_o D$ of G_{x_o} as

$$(x_o D)(ah) = x_o(a) D(h) \tag{A.6}$$

that acts on the same Hilbert space $\mathcal{K}$.

We are now ready to define the unitary representation of G *unitarily induced* by $x_o D$.

Let $\mathcal{H}$ be the Hilbert space $L^2(G[x_o], \mu, \mathcal{K})$ and fix a measurable section for the action of G on $G[x_o]$.

For each $g \in G$ we define the map U_g acting on $L^2(G[x_o], \mu, \mathcal{K})$ as

$$(U_g f)(x) := (x_o D)(c(x)^{-1} g c(g^{-1}[x])) f(g^{-1}[x]), \tag{A.7}$$

where $f \in L^2(G[x_o], \mu, \mathcal{K})$.

One has that $g \mapsto U_g$ is a unitary representation of G, which is denoted by $U = \mathrm{Ind}\,^G_{G_{x_o}}(x_o D)$.

We observe that since $g = ah$ and the action of A on $\hat{A}$ is trivial, that is, $a[x] = x$ for all $x \in \hat{A}$, we may choose the section c such that it take values on H only, that is, $c(x) \in H$ for all $x \in G[x_o]$. With this choice U takes the following form for any $g = ah$:

$$(U_{ah} f)(x) := x(a) D(c(x)^{-1} h c(h^{-1}[x])) f(h^{-1}[x]) \ . \tag{A.8}$$

The following fundamental results concerning the above construction, known as the *Mackey Machine*, are then obtained [35]:

Theorem 9. 1) *The induced representation* $\mathrm{Ind}\,^G_{G_{x_o}}(x_o D)$ *is irreducible if and only if* D *is irreducible.* 2) *Two induced representations* $\mathrm{Ind}\,^G_{G_{x_o}}(x_o D)$ *and* $\mathrm{Ind}\,^G_{G_{x_1}}(x_1 D)$ *of* G *are unitarily equivalent if and only if there is an* $h \in H$ *such that* $G_{x_o} = h G_{x_1} h^{-1}$ *and the inducing representations* $g \mapsto (x_o D_o)(hgh^{-1})$ *and* $g \mapsto (x_1 D_1)(g)$ *of* G_{x_1} *are unitarily equivalent.* 3) *Each unitary irreducible representation of* G *in a Hilbert space is equivalent to an induced one.*

References

1. V. Bargmann, On unitary ray representations of continuous groups, *Ann. Math.*, 1 (1954).
2. V. Bargmann, Note on Wigner's theorem on symmetry operations, *J. Math. Phys.* **5**, 862 (1964).
3. E. Beltrametti, G. Cassinelli, *The Logic of Quantum Mechanics*, Addison-Wesley, Reading, Massachussets (1981).
4. S.K. Bose, The Galilean group in $2+1$ space-times and its central extension, *Comm. Math. Phys.* **169**, 385 (1995).
5. S.K. Bose, Representations of the $(2+1)$-dimensional Galilean group, *J. Math. Phys.* **36**, 875 (1995).
6. J. Braconnier, Sur les groupes topologiques localements compact, *J. Math. Pure Appl.* **27**, 1 (1948).
7. F. Bruhat, Sur les représentations induites des groupes de Lie, *Bull. Soc. Math. France*, **84**, 97-205 (1956)
8. P. Busch, M. Grabowski, P. Lahti, *Operational Quantum Physics*, Lect. Notes Phys. **m31**, Springer Verlag, Berlin, Heidelberg, New York (1995), the second corrected printing 1997.
9. G. Cassinelli, E. De Vito, P. Lahti, A. Levrero, Symmetry groups in quantum mechanics and the theorem of Wigner on the symmetry transformations, *Rev. Math. Phys.* **9**, 921 (1997).
10. G. Cassinelli, E. De Vito, A. Levrero, Galilei invariant wave equations, *Rep. Math. Phys.* **43**, 467 (1999).
11. E.B. Davies, *Quantum Theory of Open Systems*, Academic Press, London (1976).
12. P.A.M. Dirac, *The Principles of Quantum Mechanics*, Oxford University Press (1930).
13. P.T. Divakaran, Symmetries and quantization: structure of the state space, *Rev. Math. Phys.* **6**, 167 (1994).
14. G.B. Folland, *A Course in Abstract Harmonic Analysis*, CRC Press, Boca Raton (1995).
15. I.M. Gel'fand, R.A. Minlos, Z.Y. Shapiro, *Representations of the Rotation and Lorentz Groups and Applications*, Pergamon Press, Oxford (1963).
16. C. George, M. Lévy-Nahas, Finite dimensional representations of non-semisimple Lie algebras, *J. Math. Phys.*, **7**, 980 (1966).
17. A.M. Gleason, Measures on the closed subspaces of a Hilbert space, *J. Math. Mech.* **6**, 567 (1957).
18. S. Gudder, P. Busch, Effects as functions on projective Hilbert space, *Lett. Math. Phys.* **47**, 329 (1999).

19. A.S. Holevo, *Probabilistic and Statistical Aspects of Quantum Theory*, North-Holand, Amsterdam (1982).
20. A.S. Holevo, *Statistical Structures of Quantum Theory*, Lect. Notes Phys. **m67**, Springer Verlag, Berlin, Heidelberg, New York (2001).
21. J. Horváth, *Topological Vector Spaces and Distributions*, vol. I, Addison-Wesley, Reading Massachusetts (1966).
22. W. Hurley, Nonrelativistic quantum mechanics for particles with arbitrary spin, *Phys. Rev. D*, **3**, 2339-2347 (1971).
23. J.M. Jauch, *Foundations of Quantum Mechanics*, Addison-Wesley (1968).
24. J.M. Lévy-Leblond, Nonrelativistic particles and wave equations, *Commun. Math. Phys.*, **6**, 286 (1967).
25. G. Ludwig, *Foundations of Quantum Mechanics*, Vol. I, Springer Verlag, Berlin, Heidelberg, New York (1983).
26. G.W. Mackey, Induced representations of locally compact groups, I, *Ann. of Math.* **55**, 101 (1952).
27. G.W. Mackey, Unitary representations of group extensions, I, *Act. Math.* **99**, 265 (1958).
28. G.W. Mackey, *Unitary Group Representations in Physics, Probability, and Number Theory*, Addison-Wesley, Reading, Massachusetts (1978, 1989).
29. L. Molnár, Z. Páles, $\perp$-order automorphisms of Hilbert space effect algebras: the two dimensional case, *J. Math. Phys.* **42**, 1907 (2001)
30. L. Molnár, Characterizations of the automorphisms of Hilbert space effect algebras, *Commun. Math. Phys.* **223** 437 (2001).
31. C.C. Moore, Extensions and low dimensional cohomology theory of locally compact groups. II, *Trans. Amer. Math. Soc.* **113**, 64 (1964).
32. C.C. Moore, Group extensions and cohomology for locally compact groups. IV, *Trans. Amer. Math. Soc.* **221**, 35 (1976).
33. L. Schwartz, *Application of Distributions to the Theory of Elementary Particles in Quantum Mechanics*, Gordon and Breach, New York (1968).
34. G. Warner, *Harmonic Analysis on Semi-Simple Lie Groups I*, Springer-Verlag, Berlin, Heidelberg, New York (1972).
35. V.S. Varadarajan, *Geometry of Quantum Theory*, second edition, Springer-Verlag, Berlin, Heidelberg, New York (1985).
36. J. von Neumann, *Mathematische Grundlagen der Quantenmechanik*, Springer, Berlin, Heidelberg, New York (1932).
37. U. Uhlhorn, Representation of symmetry transformations in quantum mechanics, *Arkiv Fysik* **23**, 307 (1962).
38. V.S. Varadarajan, *Lie Groups, Lie Algebras, and Their Representations*, Springer Verlag, Berlin, Heidelberg, New York (1984).
39. E.P. Wigner, *Gruppentheorie und ihre Anwendung auf die Quantenmechanik der Atomspektrum*, Fredrick Vieweg und Sohn, Braunschweig, Germany, 1931, pp. 251-254, *Group Theory and Its Application to the Quantum Theory of Atomic Spectra*, Academic Press Inc., New York, 1959, pp. 233-236.
40. E.P. Wigner, Unitary representations of the inhomogeneous Lorentz group, *Ann. Math.*, **40**, 149 (1939)

List of Frequently Occurring Symbols

1. Sets of Numbers

$\mathbb{N} = \{0, 1, 2, \dots\}$		74
$\mathbb{R}$	real numbers	10
$\mathbb{C}$	complex numbers	2
$\mathbb{T} = \{z \in \mathbb{C} \mid \lvert z\rvert = 1\}$		9, 97

2. Hilbert space notations

$\mathcal{H}$	complex separable Hilbert space	2, 92
$\varphi, \psi, \dots$	elements of $\mathcal{H}$	
$\langle \cdot, \cdot \rangle$	inner product of $\mathcal{H}$	2, 92
$[\varphi] = \{c\varphi \mid c \in \mathbb{C}\}$		2
$P[\varphi]$	projection on $[\varphi]$	2
$\mathbf{M} = \{M \subset \mathcal{H} \mid M \text{ closed subspace}\}$		17
$\dim(M)$	dimension of $M \in \mathbf{M}$	

3. Sets of operators on $\mathcal{H}$

$\mathbf{B}$	bounded operators	2, 90
$\mathbf{B}_r = \{A \in \mathbf{B} \mid A^* = A\}$	bounded selfadjoint operators	96
$\mathbf{B}_r^+ = \{A \in \mathbf{B}_r \mid A \geq O\}$	bounded positive operators	95
$\mathbf{B}_1$	trace class operators	2, 97
$\mathbf{B}_{1,r} = \{T \in \mathbf{B}_1 \mid T^* = T\}$		8
$\mathbf{B}_{1,r}^+ = \{T \in \mathbf{B}_{1,r} \mid T \geq O\}$		8
$\mathbf{U} = \{U \in \mathbf{B} \mid U^{-1} = U^*\}$	unitary operators	9, 98
$\overline{\mathbf{U}}$	antiunitary operators	9, 89
$\mathbf{T} = \{zI \mid z \in \mathbb{T}\}$		18
$\mathbf{S} = \{T \in \mathbf{B}_1 \mid T \geq O, \operatorname{tr}[T] = 1\}$	state operators	2
$\mathbf{E} = \{E \in \mathbf{B} \mid O \leq E \leq I\}$	effect operators	3
$\mathbf{D} = \{D \in \mathbf{B} \mid D^2 = D^* = D\}$	projection operators	3, 95
$\mathbf{P} = \{P \in \mathbf{D} \mid \dim P(\mathcal{H}) = 1\}$		2, 95

4. Groups of automorphisms

$\mathrm{Aut}\,(\mathbf{S})$	state automorphisms	8
$\mathrm{Aut}_s(\mathbf{P})$	superposition automorphisms	10
$\mathrm{Aut}\,(\mathbf{P})$	vector state automorphisms	10
$\mathrm{Aut}_0(\mathbf{P})$	transition probability zero preserving bijective functions $p : \mathbf{P} \to \mathbf{P}$	10
$\mathrm{Aut}(\mathbf{E})$	effect automorphisms	15

$\mathrm{Aut}_o(\mathbf{E})$	effect $\perp$-order automorphisms	12
$\mathrm{Aut}_s(\mathbf{E})$	effect sum automorphisms	12
$\mathrm{Aut}_c(\mathbf{E})$	effect convex automorphisms	12
$\mathrm{Aut}\,(\mathbf{D})$	$\mathbf{D}$-automorphisms	14
$\mathrm{Aut}\,(\mathcal{H}) = \mathbf{U} \cup \overline{\mathbf{U}}$		18, 99
$\Sigma = \mathbf{U} \cup \overline{\mathbf{U}}/\mathbf{T}$		18, 99
$\Sigma_0 = \mathbf{U}/\mathbf{T}$		99
$\Sigma(\mathcal{H}, \mathcal{H}')$		28

5. Some mappings

$\pi : \mathbf{U} \cup \overline{\mathbf{U}} \to \Sigma$	canonical projection	25
$\pi : \mathbf{U} \to \Sigma_0$	canonical projection	29
$s : \Sigma \to \mathbf{U} \cup \overline{\mathbf{U}}$	a section for $\pi : \mathbf{U} \cup \overline{\mathbf{U}} \to \Sigma$	25
$s : \Sigma_0 \to \mathbf{U}$	a section for $\pi : \mathbf{U} \to \Sigma_0$	29
$\sigma : G \to \Sigma$	a symmetry action	28
$\sigma : G \to \Sigma_0$	a (unitary) symmetry action	29
$\tau : H \times H \to A$	an A-multiplier of H	31
$z : G \times G \to \mathbb{T}$	a $\mathbb{T}$-multiplier of G	32
$\tau_F : H \times H \to \mathbb{R}^n$	an $\mathbb{R}^n$-multiplier of H associated with a closed $\mathbb{R}^n$-form F	33
$F : \mathrm{Lie}\,(H) \times \mathrm{Lie}\,(H) \to \mathbb{R}^n$	a closed $\mathbb{R}^n$-form	32
$\delta : G^* \to G$	covering homomorphism	33
$\rho : \overline{G} \to G$	$\rho(v, g^*) := \delta(g^*)$	34
$c : G \to \overline{G}$	a section for ρ	37

6. Groups

G	a symmetry group	30
G^*	covering group of G	30
$\overline{G}$	universal central extension of G	34
$\overline{G} = A \times' H$	semidirect product	43
A	normal closed Abelian subgroup of $\overline{G}$	43
H	closed subgroup of $\overline{G}$	43
$g = (a, h) \in \overline{G}$	an element of $\overline{G}$	43
$\hat{A}$	dual group of A	43
$\overline{G}_x$	stability subgroup of $\overline{G}$ at $x \in \hat{A}$	43
$\overline{G}[x]$	orbit corresponding to $\overline{G}_x$	43
ν	$\overline{G}$-invariant σ-finite measure on $\overline{G}[x]$	43
$\mathcal{K}$	Hilbert space	43
$L^2(\overline{G}[x], \nu, \mathcal{K})$	Hilbert space of functions	43
D	representation of $\overline{G}_x \cap H$ acting in $\mathcal{K}$	43
xD	representation of $\overline{G}_x$ defined by $(xD)_{ah} = x_a D_h$	43
$U = \mathrm{Ind}_{\overline{G}_x}^{\overline{G}}(xD)$	representation of $\overline{G}$ induced by xD	43

$H^2(G, \mathbb{T})$	quotient group corresponding to the group of $\mathbb{T}$-multipliers of G	32
$H^2(G^*, \mathbb{R})$	vector space of the equivalence classes of the $\mathbb{R}$-multipliers of G^*	34
$H^2(G^*, \mathbb{R})_\delta$	a subspace of $H^2(G^*, \mathbb{R})$	34
$\text{Lie}\,(H)$	Lie algebra of H	32
$H^2(\text{Lie}\,(H), \mathbb{R}^n)$	quotient space corresponding to the set of closed $\mathbb{R}^n$-forms	33

7. Notations related to the Galilei group in $3+1$ dimensions

$\mathcal{V} := (\mathbb{R}^3, +)$	velocity transformations	49
$SO(3)$	rotations in $\mathbb{R}^3$	49
$G_o := \mathcal{V} \times' SO(3)$	homogenous Galilei group	49
$(\boldsymbol{v}, R) \in G_0$	an element of G_0	
$SU(2)$	covering group of $SO(3)$	50
$G_0^* = \mathcal{V} \times' SU(2)$	covering group of G_0	
$(\boldsymbol{v}, h) \in G_0^*$	an element of G_0^*	
$\mathcal{T}_s := (\mathbb{R}^3, +)$	space translations	49
$\mathcal{T}_t := (\mathbb{R}, +)$	time translations	49
$\mathcal{T} := \mathcal{T}_s \times \mathcal{T}_t$		
$(\boldsymbol{a}, b) \in \mathcal{T}$	an element of $\mathcal{T}$	
$G := \mathcal{T} \times' G_o$	Galilei group	49
$g = (\boldsymbol{a}, b, \boldsymbol{v}, R)$	a Galilei transformation	50
$G^* = \mathcal{T} \times' G_o^*$	covering group of G	50
$g^* = (\boldsymbol{a}, b, \boldsymbol{v}, h)$	an element of G^*	
$\delta((\boldsymbol{a}, b, \boldsymbol{v}, h)) = (\boldsymbol{a}, b, \boldsymbol{v}, \delta(h))$	covering homomorphism $G^* \to G$	50
$\overline{G} = \mathbb{R}^5 \times' (\mathcal{V} \times' SU(2))$	universal central extension of G	55
$\bar{g} = (\boldsymbol{a}, b, c, \boldsymbol{v}, h)$	an element of $\overline{G}$	65
$\text{Lie}\,(\mathcal{T}) = \mathbb{R}^4$	Lie algebra of $\mathcal{T}$	51
$\text{Lie}\,(\mathcal{V}) = \mathbb{R}^3$	Lie algebra of $\mathcal{V}$	51
$\text{Lie}\,(SO(3)) = so\,(3)$	Lie algebra of $SO(3)$	51
$\text{Lie}\,(G_0) = \mathbb{R}^3 \oplus so\,(3)$	Lie algebra of G_0	51
$\text{Lie}\,(G) = \mathbb{R}^4 \oplus \mathbb{R}^3 \oplus so\,(3)$	Lie algebra of G	51
$\text{Lie}\,(G^*) = \mathbb{R}^4 \oplus \mathbb{R}^3 \oplus su\,(2)$	Lie algebra of G^*	52
$\text{Lie}\,(\overline{G}) = \mathbb{R} \oplus (\text{Lie}\,(\mathcal{T}) \oplus (\text{Lie}\,(\mathcal{V}) \oplus su\,(2)))$	Lie algebra of $\overline{G}$	54

8. Miscellaneous

$D_F(t_1, t_2)$	the time evolution operator of the frame F	62
$\mathbb{P}^n$	the dual group of the additive group $\mathbb{R}^n$, $n = 2, 3, 4, 5$	65
$\leq$	order on $\mathbf{B}_r$	
$\perp$	$\mathbf{E} \ni E \mapsto E^\perp := I - E \in \mathbf{E}$	
$\vee$	least upper bound w. r. t. $\leq$	
$\wedge$	greatest lower bound w. r. t. $\leq$	

Index

Lecture Notes in Physics

For information about Vols. 1–607
please contact your bookseller or Springer
LNP Online archive: springerlink.com

Printing: Strauss GmbH, Mörlenbach
Binding: Schäffer, Grünstadt